AF706060

THE RELIABILITY FRAMEWORK

Preventive Maintenance

THE RELIABILITY FRAMEWORK • PART 5

5

Preventive Maintenance

Design and Optimize
Your Maintenance Schedule

MARTIN VAN DEN HOUT

VULCANUS PRESS
FABRICANDO FIT FABER

Colophon

Title:	Preventive Maintenance: Design and Optimize Your Maintenance Schedule
Series:	The Reliability Framework: Nine Guides to Flawless Asset Performance
Part:	5
Author:	Martin van den Hout
Photo’s:	Depositphotos.com
ISBN:	9789083643168
Edition:	First edition: April 2026

Liability and Disclaimer

The information in this book has been compiled with great care, drawing on 35 years of practical experience in maintenance and reliability engineering. The goal is to provide maintenance professionals with practical tools for improving their maintenance processes.
Neither the author Martin van den Hout, nor Vadeo BV, nor Vulcanus Press can accept liability for the consequences of actions based on this book. Every business situation is unique and requires its own assessment of risks, costs, and benefits.

When implementing the described methods, you should take into account the specific circumstances of your organization, applicable laws and regulations, safety requirements, and manufacturer guidelines. When in doubt, consult specialists and make responsible decisions based on thorough risk analysis.

This book is intended as a practical guide and does not replace customized professional advice, adequate employee training, or sound engineering judgment.

Foreword

Anyone who works daily to keep their assets reliable knows how exceptional that profession is. Behind every running pump, every production line that keeps operating, and every network that keeps functioning, stands a skilled professional with knowledge, dedication, and an unstoppable curiosity for improvement. The maintenance professional is the silent force that makes technology perform, monitors processes, and moves organizations forward — often without anyone noticing, until something goes wrong.

Yet the world of maintenance is changing faster than ever. From TPM to Lean, from Asset Management to Reliability Engineering – the terminology shifts, but the core remains the same: reliability starts with people who know what they're doing, and who dare to learn, share, and improve.

This book is an invitation to you – the professional who looks beyond repairing what is broken. It is a compass for those who view maintenance not as a cost, but as the engine of continuity, safety, and value creation. Whether you work in industry, the public sector, or at a high-tech company, the principles shared here are universal: structured thinking, craftsmanship, and working together toward reliability.

Let this book inspire you to look beyond today. See every maintenance moment as an opportunity to learn, every problem as a reason to improve, and every asset as a system that can do better — because you want to be better.

My admiration goes out to Martin van den Hout, author and knowledge custodian with decades of experience in the maintenance and Asset Management field.

His sharp insights, thorough knowledge, and passion for sharing expertise have made an invaluable contribution to this work. Thanks to his ability to make complex subject matter clear and accessible, his contributions form a valuable bridge between theory and practice.

Because, ultimately, maintenance is not just a profession: It is a mindset!

Ellen den Broeder-Ooijevaar
Manager NVDO
(Dutch National Maintenance Society)

Table of Contents

Introduction

This book is part of a series that answers the question: "What should an organization do to make its technical installations perform as well as possible?"
I asked myself this question 35 years ago, and it has kept me constantly engaged ever since. At the time, I was working in the Maintenance & Engineering department of a Japanese manufacturing company in the Netherlands. In those days, the term "Total Productive Maintenance" (TPM) was on the rise, and more and more people were beginning to realize that simply setting up a good maintenance schedule is not enough to keep a production plant running smoothly.
Within TPM, the emphasis was strongly on involving production operators and improvements within a production department. I realized fairly quickly that this was a major step forward, but still not enough.

There were also many questions within our maintenance department that TPM didn't answer. Questions like: "Which spare parts do we need, and how many?" or "What is the best way to organize the large flow of work within a maintenance department, ranging from jobs that take a minute to projects requiring thousands of man-hours?"
In the past thirty-five years, our field has matured. TPM was incorporated into Lean Manufacturing, and from (semi-) government bodies, more and more attention was paid to technical management, under the name asset management. This ultimately led to the publication of the ISO 55000 series of standards. For the reliable operation of a production plant or other facility, the term 'Reliability' emerged as a replacement for the term 'Maintenance'. After all, reliability is the goal, while maintenance is merely one of the means.

I learned that all these schools of thought strongly resembled one another and that the differences were mainly differences in emphasis, driven by corporate culture and factors such as how many operators are directly involved in production. In the chemical industry their role is different from that in the automotive industry, and in government organizations this role is different again.

Ultimately, many organizations have combined elements from these approaches in whatever way best suited their specific situation. Yet I learned that what applies to a production plant also applies to a network, your car, or a power plant. It doesn't matter.

It always bothered me that there were only two kinds of books about this field. The first kind describes the entire field, but only describes 'what' you should organize. They are therefore very broad, but never go into any depth. The second kind focuses on one aspect, such as workflow control or setting up preventive maintenance, and does go into depth on how to apply that one aspect.
These books have a number of things in common. They are not interconnected with one another, there are very few good ones, they are usually written in foreign languages, and they are always very thick. Often they are also written from the perspective of 'the ideal world,' in which you have an infinite number of engineers or technicians at your disposal and the most fantastic computer programs.

How This Book Is Structured

This book deliberately follows the natural steps you go through when setting up an effective preventive maintenance program. You can think of it as a journey in three phases, where each chapter builds on the previous one.

The Foundations (Chapters 1–4) begins with the why and the what.
Chapter 1 explains why a systematic approach to preventive maintenance is essential and what economic reality lies behind it. You will learn about the concrete benefits and understand why sustainability makes preventive maintenance even more relevant.

Chapter 2 covers establishing clear requirements — perhaps the most important step in the entire process. Here you learn how to work with management to define concrete availability requirements, map out risks, and lay the foundation for all further decisions. These requirements determine your maintenance strategy.

In chapter 3 you will gain practical techniques for inventorying all your assets. From setting up a complete asset register to efficiently gathering information without getting bogged down in details. This inventory forms the basis for all further analyses.

Chapter 4 shows you how to systematically determine the criticality of your assets. With the criticality ranking methodology, you learn within a few days which assets deserve the highest priority and which maintenance methodology fits each category.

From analysis to strategy (Chapters 5–7) is where theory becomes practice.
Chapter 5 forms the heart of the book: here you learn three methodologies for determining preventive maintenance. From full RCM analyses for critical assets, through Industrial RCM for important assets, to pragmatic approaches for standard assets. Sarah's complete RCM analysis of pump P-401 shows step by step how you apply this in practice.

Chapter 6 goes in depth on why assets fail. The six failure patterns, technical aging mechanisms, and human factors give you the insight needed to choose effective preventive measures. This knowledge makes the difference between maintenance that works and maintenance that is waste.

In chapter 7, maintenance types become practical. From usage-based via condition-based to function tests and modern IoT monitoring. You will learn when each technique is most effective and how to implement it in your specific situation.

From Plan to Practice (Chapters 8–10) ensures that your program doesn't become a paper tiger.
Chapter 8 makes economic optimization practical. You learn to weigh costs and benefits, determine optimal intervals, and substantiate investment decisions. The sample calculations give you tools you can apply immediately.

Chapter 9 describes the practical implementation with all the challenges that Sarah encountered at TechPro. From drafting workable work orders to ensuring execution. The 6-month roadmap shows how to systematically scale up from pilot to company-wide implementation.

Digital Amplification and the Future (Chapter 10) looks ahead.
Chapter 10 covers how IoT sensors, AI, and cloud platforms can strengthen your RCM methodology. The message is clear: digitalization reinforces systematic thinking, but does not replace systematic thinking. The epilogue invites you to move on to the complete methodology of nine disciplines.

Each chapter concludes with a natural transition to the next, so that you can see how all the elements relate to each other. The ongoing case study of Sarah and TechPro makes abstract concepts concrete and recognizable. In the appendices you can find practical tools and templates that you can use immediately.

This series consists of nine books, which together provide the answer to the previously mentioned question: "What should an organization do to make its technical installations perform as well as possible?" The books cover one topic each in a systematic, step-by-step approach. Each book connects to the other books in the series. This way, you can implement a coherent system within your company to allow your assets to perform optimally.

This is the fifth book in this series. The series consists of the following books:

Part 1: Maintenance Management

A comprehensive overview of the entire methodology. A general description of the topics covered in the other eight parts, plus several additional subjects.
ISBN 9789083643120

Part 2: Planning and Scheduling in Maintenance

Organization of the workflow of the maintenance department; including work requests, gate-keeping, work preparation, scheduling, supervision, safety, administration, improvement, and archiving.
ISBN 9789083643137

Part 3: Troubleshooting Breakdowns

The resolution of an individual failure, by a maintenance technician or an operator. Notification, finding the cause, finding a solution, repair, failure reporting, continuous improvement.
ISBN 9789083643144

Part 4: Root Cause Analysis in Maintenance

Reducing the number of failures that occur annually. Registration, statistical analysis, prioritization, root cause analysis, and implementation of solutions.
ISBN 9789083643151

Part 5: Preventive Maintenance

Setting up a preventive maintenance schedule for an entire organization, taking into account the risks and criticality of the individual assets. Translating management requirements for the entire organization into technical requirements for the assets. Mapping possible failure modes, causes, and possible maintenance actions. Economic approaches to determine which maintenance is worthwhile. Methodologies such as RCM, Industrial RCM, and pragmatic approaches.
ISBN 9789083643168

Part 6: Spare Parts Management

What types of spare parts are there? Is it worthwhile to keep them in stock? How many of each type? Organization of the warehouse and key considerations for common types of components.
ISBN 9789083643175

Part 7: Precision Maintenance

Many assets fail prematurely because they have not been assembled with sufficient precision. A technical handbook covering practical topics such as bolt tightening, fits, aligning rotating equipment, assembly with attention to balance, flange assembly, and so forth.
ISBN 9789083643182

Part 8: Engineering for Reliability

80 percent of maintenance costs and technical reliability is determined during design and construction. Project management, Early Equipment Management, involvement of operators and technicians, reliability analyses during design, FAT and SAT, Commissioning, Maintainability.
ISBN 9789083643199

Part 9: Autonomous Maintenance

Which maintenance tasks can you have operators perform, why, and how do you organize that?
ISBN 9789083706306

For a current overview of all books, additional information, and ordering options, visit:

vadeo.nl/books.html

1. Preventive Maintenance – The Business Case

Sarah Shoemaker, reliability engineer at TechPro Manufacturing, stared at the maintenance costs from the past quarter. Nearly 40% of all work orders were unplanned repairs. Machines failing unexpectedly, production lines standing idle, technicians rushing to replace parts. The costs were mounting, delivery reliability to customers was declining, and management wanted answers. "There has to be a better way," she thought.

Sarah was right. This book shows you the systematic approach she discovered to transform TechPro from a reactive to a preventive maintenance organization. But before you dive into the methodology, it is important to understand why this transformation is not only technically sound, but also has a solid business case.

1.1. The Economic Reality

Maintenance represents a significant cost item for most capital-intensive organizations. At TechPro, it ranked among the top three operational expenditures, and this pattern is not unusual. This scale regularly leads to cost-cutting pressure from management, especially when budgets are under strain.

Here, however, lies a dangerous pitfall. Ill-considered cuts in maintenance work counterproductively. They lead to more failures that still need to be repaired, often at considerably higher costs than preventive maintenance would have cost. An unplanned failure requires not only emergency measures and rush deliveries of parts, but also causes production downtime, missed deadlines, and potential quality issues.

The economically optimal balance lies not in minimal maintenance, but in smart maintenance. One invests preventively where it is most effective, and consciously accepts certain risks where the cost of preventive maintenance exceeds the expected damage. This balance differs per asset, per business situation, and per risk profile.

1.2. Proven Results

Organizations that systematically implement preventive maintenance consistently achieve the following improvements:

30–50% reduction in unplanned downtime – The most direct impact. Failures are not eliminated, but their frequency drops dramatically.

20-30% reduction in total maintenance costs – This may seem paradoxical, but it results from more efficient use of resources and avoiding costly emergency repairs.

15–25% improvement in Overall Equipment Effectiveness – Machines run more stably, with fewer minor failures and disruptions.

ROI of 150–300% within 18 months – Investments in systematic approaches pay for themselves quickly through lower costs and higher productivity.

For Sarah, these numbers meant that TechPro's investment in a systematic approach was not a cost item, but a strategic investment with measurable returns. Management would see not only cost reduction, but also improved delivery reliability to customers.

1.3. Sustainability as an Additional Driver

Over the last few years, a new factor has become increasingly important: sustainability. Where this was once primarily a "nice to have" for corporate social responsibility reports, it has now become a hard business driver. Customers place demands on the CO2 footprints of their suppliers. Regulations are becoming stricter. And energy prices are making efficiency an economic necessity. Preventive maintenance directly contributes to these sustainability objectives:

Well-maintained assets consume less energy.

A pump installation with properly aligned shafts, clean filters, and optimal lubrication consumes 5–15% less energy than a poorly maintained equivalent. Over thousands of operating hours and dozens of assets, these percentages add up quickly.

Assets last longer.

A machine that is systematically maintained reaches its design life or even exceeds it. This means less frequent replacement and therefore less use of raw materials and energy for new machines.

Fewer spare parts and waste.

Preventive maintenance prevents not only major failures, but also their cascading effects. An undetected leaking gasket can lead to bearing damage, oil contamination, and ultimately to the replacement of far more components than just that gasket.

1.4. Compliance and Documentation

Sarah discovered a third driver that management often overlooks: the growing requirement to be able to justify your maintenance approach. Whether you work under ISO 9001, ISO 13485 for medical equipment, SOLAS for maritime shipping, or other industry-specific safety standards – modern business operations require that you can demonstrate your maintenance system actually functions as intended.

What Auditors and Inspectors Require

The specific standards differ, but the basic requirements are universal:

Prove that you know which assets are critical to your operations and what risks they carry
Show that you have adequate maintenance plans that address these risks
Document that you are carrying out these plans as intended
Keep records of all performed activities
Ensure that qualified personnel work with calibrated equipment

At TechPro, this hit hard during an ISO audit. The auditor asked about the maintenance plan for a critical chiller. "We follow the manufacturer's specifications," was the answer. "Can you demonstrate that these specifications are adequate for your specific situation?" was the follow-up question. TechPro could not. The observation in the audit report was clear: no systematic risk-based approach to maintenance.

1.5. The Business Impact

This documentation requirement is not bureaucratic red tape. It has direct business consequences:

Customers are increasingly choosing suppliers who can prove their quality and reliability. Especially in regulated sectors such as automotive, aerospace, and medical devices, this has become a hard selection criterion.
Insurance companies charge lower premiums for companies that can demonstrably show good maintenance practices. TechPro's insurer had already announced that companies without a documented maintenance program would face a premium increase at the next renewal.

Regulatory agencies can impose severe penalties on organizations that cannot document their maintenance obligations. In some sectors, this can even lead to temporary shutdown of operations.

The Good News

A systematic approach to preventive maintenance automatically generates the documentation that external parties require. Your criticality

analysis becomes your risk documentation. Your methodical selection of maintenance strategies produces well-substantiated plans. Your structured execution creates the audit trail that proves compliance.
For Sarah, this meant that TechPro's new maintenance approach would kill three birds with one stone: operational improvement, cost reduction, and compliance with all relevant standards.

1.6. Sarah's Assignment and Approach

With this business case, Sarah presented her proposal to management. The cost-benefit analysis was compelling, the sustainability advantages supported the business strategy, and the compliance aspects could no longer be ignored. She was given 6 months to transform TechPro from 70% reactive to 70% preventive maintenance.

Sarah discovered a methodology consisting of 8 logical steps:

1	Define clear requirements	What should the installation be capable of? What risks are acceptable?
2	Inventory all assets	You Can Only Manage What You Know
3	Determine the Criticality	Not all assets deserve the same attention
4	Choose the Right Methodology	RCM for Critical Assets pragmatic for standard assets, Industrial RCM for the group in between
5	Analyze Failure Modes	Why Do Assets Fail and What Can You Do About It?
6	Select Maintenance Types	From Inspections to Condition-monitoring
7	Optimize Economically	What is the optimal investment?
8	Implement and Sustain	From Plan to Practice

These 8 steps form the backbone of this book. Each chapter takes you through one step, with practical tools and examples from Sarah's experience at TechPro.

1.7. Your Situation

Maybe you work at a manufacturing company like TechPro. Or maybe you manage a building management system, a water transport network, or a fleet of vehicles. The assets differ, but the methodology remains the same. All technical systems share similar challenges: they wear out, they can fail, and the question is always how to find the optimal balance between reliability and cost.

This book gives you the tools to find that balance in your specific situation. The methodology has been developed and tested across diverse industries, from chemical plants to hospitals, from offshore platforms to data centers. The principles are universal, the application specific.

1.8. The Journey Begins

Chapter 2 begins where Sarah began: with defining clear requirements. Before you can determine what maintenance your assets need, you must know what you expect from those assets. What availability is required? What risks are acceptable? What are the consequences of failure?

These are the fundamental questions that shape your maintenance plan. Sarah discovered that many of TechPro's maintenance problems stemmed from unclear or unspoken expectations. Management wanted "maximum reliability," but had never defined what that meant. Operators wanted "stable production," but had never quantified what was acceptable.

Without clear requirements, every maintenance program is a shot in the dark. With clear requirements, you can systematically build a program that delivers exactly what your organization needs.

Let's get started.

2. Requirements for Effective Maintenance

Sarah stared at the conference table where director Harrison and the production managers had taken their seats. "We need to get specific," she began. "What happens if our main compressor fails during peak production?" The answer came quickly: $3,500 per hour in production loss, plus a possible $12,000 rejected batch. These concrete figures suddenly made the budget for preventive maintenance make sense.

As indicated in the introduction, clearly defining requirements forms the foundation of every effective maintenance plan. Before you can develop a maintenance plan, you first need to know exactly what requirements a technical system must meet. This chapter covers the various aspects involved in this process.

Without clear requirements, you run the risk of performing maintenance that is technically excellent but does not align with what your organization truly needs. You might, for example, perform perfect, comprehensive maintenance actions on a pump, but if that pump is not at all important to your organization, you are probably doing too much and the company is wasting time and money.

2.1. System Requirements and Performance Standards

Every organization places different demands on its technical systems, and even within your organization these demands can vary from asset to asset. For some production lines, continuous availability is critical due to high product demand, while other lines can sit idle for weeks without significant consequences. In certain situations, a failure can lead to the complete rejection of a production batch with considerable financial damage, while in other cases the impact remains limited to a brief production interruption.

It is therefore essential that you establish the exact functional requirements and performance standards for each asset. Take as a simple example the lighting in a workspace. You can set different requirements: must all lights be functioning, is it sufficient if a minimum percentage is functioning, or is a specific light intensity (in lux) required at each workstation?

This seemingly simple question illustrates a fundamental principle: the requirements you set directly determine what maintenance is necessary. If you require all lights to function, you must replace defective bulbs immediately. If you require a minimum light intensity, you can wait until multiple bulbs have failed before taking action.

2.2. Practical Approach to Requirements Definition

In practice, managers often struggle to clearly formulate these requirements, especially when they lack a technical background. It is therefore advisable to organize a structured management session in which you explain why clear requirements are necessary and ask targeted questions to identify these requirements.

This management session has two important goals. First, you obtain clear, measurable requirements for each technical system. Second, you create awareness among management regarding the potential cost savings and productivity improvements that can be achieved through systematic maintenance. This management session forms step 1 of the 8-step methodology and lays the foundation for all subsequent phases.

Many managers view maintenance primarily as a cost and fail to understand that well-designed maintenance is an investment that pays for itself through reduced failures and increased productivity. It is important that you help them see that maintenance is a form of engineering: if the requirements are clear and the behavior of the assets is known, you can as an engineer develop an optimal maintenance plan that meets the needs within the available budget.

An effective method is to discuss concrete examples from the organization itself during this management session. Ask, for example: "What is the maximum amount of time our production can be down?" or "How much money may we spend to avoid losing one hour of production capacity to failures?" These specific questions help managers formulate concrete requirements instead of vague wishes.

Sarah's management session at TechPro was a breakthrough. "I had prepared three concrete scenarios in advance," she said later. "Failure of the main compressor, a failure in the cooling water system, and problems with the packaging line. By discussing these concrete cases, we finally got clear numbers: $3,500 per hour for the compressor, $1,800 per hour for cooling, and $800 per hour for packaging."

2.3. Availability and Reliability

A technical system is acquired or built to fulfill a specific function. A packaging machine packages products, a welding machine welds, and a pump system transports liquids. To define availability requirements, you can ask the following questions: what percentage of the time must the system be available, is the system fully scheduled or is there overcapacity, and what is the maximum acceptable duration of an unplanned downtime?

The Importance of Failure Duration

This last question is particularly important. A failure is always undesirable, because the company generates no revenue during that time, while costs continue. But not every failure is equally critical. For some systems, failures only become truly problematic when they last too long — for example, because customers may then switch to another supplier, or because a prolonged interruption can compromise a multi-year research project.

You also need to discuss the frequency and duration of failures. Pose this question: "What is worse: ten short failures or one long failure?" The answer depends on the specific process. In some processes, short failures lead to major problems, for example when restarting after a failure takes many hours or when an entire production batch has to be rejected. In that situation, ten short failures would be absolutely catastrophic. In other cases, such as a logistics center that promises 24-hour delivery, longer failures in particular can disrupt the entire business process.

These differences have direct consequences for your maintenance strategy. If short failures are very costly, you invest more in continuous monitoring to get early warnings. If long failures are the problem, you focus on measures that prevent catastrophic breakdowns, even if you accept more minor interruptions in the process.

Reliability versus Availability

It is important that you understand the distinction between reliability and availability. Reliability indicates how often a system fails, while availability indicates what percentage of the time the system is operational. A system can have high availability through fast repairs, even if it fails regularly. Conversely, a highly reliable system can have low availability if repairs take a long time. For your maintenance strategy, this means you must choose the balance between preventing failures (increasing reliability) or reducing repair times (increasing availability). This choice depends on your specific operational requirements and economic considerations.

2.4. Costs and Economic Factors

It is not realistic to strive for 100% failure-free operation of all systems. Some failures occur so rarely that preventive maintenance to prevent them is not economically justifiable. The crucial question is therefore: how much money may you spend to prevent a failure of a given duration? This question is more specific and more useful than "What does one hour of downtime cost?", because the latter phrasing often leads to endless discussions depending on the perspective. As a reliability engineer, you want to know how much budget you are allowed to spend on preventive measures to avoid a specific failure.

2.5. Overcapacity and Schedulable Production

An important consideration here is whether the system has overcapacity. A system that can produce more than strictly necessary can fail occasionally without serious consequences. Some companies therefore do not aim for maximum production, but for predictable and planned production.

The costs of downtime can vary depending on the duration and timing. The first 30 minutes of downtime per shift may cost little, because this backlog can be caught up within the same shift. Up to two hours of downtime per week can perhaps be compensated within the same week. Longer downtime may necessitate extra shifts on the weekend.

In addition to production loss, you should also ask about the costs of a rejected product or a rejected production batch, and the costs of delivery delays and any penalty clauses.
This layered cost estimation helps you establish realistic budgets for different types of preventive maintenance. For example, a measure that prevents a four-hour shutdown may cost more than a measure that only prevents shutdowns of 30 minutes.

2.6. Risk Management

Strictly speaking, the risk of an event is the probability that the event occurs multiplied by the consequence of that event. An event with a low probability and a large consequence can therefore have the same risk as an event with a high probability and a small consequence.
This definition makes clear why risks require a different approach than purely economic considerations. With economic impact, you seek an optimal balance between preventive maintenance costs and the costs of failures. With risks whose consequences cannot easily be expressed in monetary terms — such as safety risks for employees or environmental contamination — you must reduce the risk to an acceptable level, where costs are secondary.

Organizations must consider a broad spectrum of risks. A comprehensive checklist of possible risk types:

- Safety – accidents, exposure to hazardous substances

- Availability — downtime, production losses
- Physical security - unauthorized access, sabotage
- Cybersecurity - digital attacks, data loss
- Health and Occupational Hygiene – long-term effects on employees
- Environment – pollution, emissions, waste
- Economic consequences – direct financial losses
- Political consequences – government measures, loss of permits
- Quality risks – product quality, service delivery
- Food safety – specific to the food industry
- Reputational damage and image damage
- Delivery reliability – to customers
- Comfort level – for users
- Staff satisfaction – motivation, absenteeism
- Minor regulatory violations – acceptance of regular violations
- Extended downtime — strategic impact

Note: This checklist can never be complete. Every organization must determine for its specific situation which risk types are relevant.
But what is an "acceptable" risk? This is a question that must be answered by management — not by the technical manager, but by senior management, which bears responsibility to government and shareholders.

What one organization finds acceptable, another may find completely unacceptable. Some managers may say they want to eliminate every risk, but that is not realistic. In every production plant or comparable facility, there is always a certain level of risk, such as a wet floor on which someone can slip. As a maintenance department, you must focus on the most significant risks.

Score	Economics	Safety	Condition	Energy consump-tion	Environ-ment	Heard of it	<1x per 10 years	<1x per year	<1x per month	>1x per month
1	< $1,100	No impact	No impact	Optimal	No conse-quences					
2	< $5,500	First aid	Minor complaint	< 5% increase	Internal spill					
3	< $55,000	Lost-time accident	Return delivery to customer	< 10% increase	Reportable incident					
4	< $550,000	Permanent disability		> 10% increase	Off-site damage					
5	> $550,000	Fatality	Product recall		Permanent off-site damage					

Risks in this category require immediate action

Risks in this category require planned action. These must be weighed against each other and prioritized

Risks in this category are acceptable

Figure 2-1 Example of a Risk Matrix

Risk Matrix as a Tool

An effective risk matrix helps with systematically assessing all relevant risk types. This matrix contains:

- Different risk types (economic, safety, environmental, quality, etc.) in the columns
- Severity levels (for example 1–5 or 1–10) on the rows
- At each severity level, a clear description of what it means
- For each combination of severity and type: how often such an incident should occur at most

Organizations use the risk matrix for various decisions:

- Replacement investments: Can deterioration lead to unacceptable risks?
- Preventive maintenance schedules: Does maintenance reduce risks to an acceptable level?
- Prioritization of improvement actions following problem occurrence
- Design new installations: where is thorough risk analysis needed?

Critical warning: Managers indicate in the risk matrix which incident frequency is acceptable for their entire area of responsibility. If they consider one lost-time accident every three years acceptable for their entire plant, this does NOT mean that in your risk analysis you may assign one accident every three years per machine.
After establishing this matrix, you can as a technical team identify the most severe potential failures and verify whether the proposed maintenance plan sufficiently mitigates these risks.

2.7. Sustainability as an Emerging Risk Area

Many organizations today also have sustainability objectives: reduction of greenhouse gas emissions, circular production processes, reduction of raw material usage, and energy savings.
These objectives have direct implications for your maintenance plan. It means that you must maintain assets in such a way that they produce minimal CO2 emissions and operate energy-efficiently, which requires regular monitoring. Raw materials are becoming increasingly scarce, and energy costs are among the highest operational expenses at many companies.

Practically, this means you must monitor energy consumption as part of your condition monitoring, take into account the environmental impact of lubricants and other maintenance materials, and prioritize asset life extension over replacement where possible.
At TechPro, sustainability proved to be an additional motivation for Sarah's preventive maintenance approach. "Our well-maintained main compressor uses 12% less energy than before the last maintenance improvement," she reported to management. "That saves not only $18,000 per year in energy costs, but also contributes to our CO2 reduction target."

From Requirements to Action

Now that you have learned how to define clear requirements for your technical systems, the next chapter will show you how to get a complete overview of all assets in your organization. After all, without knowing what you have, you cannot create an effective maintenance plan. The next chapter teaches you practical techniques for setting up an asset register and how to work efficiently without getting bogged down in unnecessary details.

3. Inventorying Assets

Sarah started the TechPro asset inventory optimistically, but quickly discovered that the asset register contained 30% incomplete entries. New machines had not been registered, replaced parts had inconsistent numbers, and assets had disappeared during the last renovation without being recorded. "No wonder our maintenance is chaotic," she realized. "How can you properly maintain something when you don't even know what you have?"

If you don't know what you have, you can't maintain it. So before you can move forward with setting up or optimizing your preventive maintenance schedule, you need a complete list of all assets in your organization. The term ISO 55000 uses for this list is "asset register." Some companies also call it a Master Equipment List (MEL).

Now it may seem obvious and easy to make such a list, and you might expect that every company already has one. In practice, however, most organizations do not have a correct and up-to-date list. In my 35 years of experience, I have worked with many companies and government organizations, and most of them have a list that comes close to reality, but still contains errors and outdated information, with a number of assets missing entirely. Some other organizations are completely off the mark and have nothing at all, or perhaps only a few items recorded.

3.1. The Importance of a Complete Asset Register

Before you can determine what maintenance your assets require, you need to know which assets you actually have. This sounds logical, but in practice this turns out to be one of the greatest challenges when setting up preventive maintenance.

A complete asset register forms the foundation for all further steps in your maintenance optimization. Without this foundation, you cannot create a realistic plan, cannot prepare accurate cost estimates, and you risk overlooking critical assets.

The problem often arises because assets are added, replaced, or removed over the years without the administration being consistently updated. Newly installed equipment is forgotten to be added, replaced parts receive new numbers that are not consistent with the existing system, and during renovations assets sometimes disappear without this being recorded.

Why an Asset Register Is Often Missing or Incomplete

There are several reasons why organizations struggle with complete asset registers.
First, its value is often underestimated. Many people see it as administrative red tape rather than as an essential foundation for effective maintenance.
Second, creating and maintaining an asset register requires discipline and consistency over long periods of time. Every time something changes in the installation, the register must be updated. In the day-to-day pressure of operational activities, this often gets low priority.
A third issue is that clarity is often lacking about what does and does not belong in the register. Should every bolt be recorded, or only main components? Where do you draw the line between assets and consumables? This lack of clarity leads to inconsistent registration.

Sarah's experience at TechPro illustrates this well. "We had three different lists," Sarah said. "One in the ERP system for financial depreciation, one in the workshop with handwritten notes, and one in the process computer. Not one of them was complete, and they were all really different."

3.2. Practical Approach to Inventory Assessment

Setting up a complete asset register requires considerable effort in many cases. You need to go through all available documents such as manuals, quotations, and installation reports to find out what was actually delivered. Then you need to verify this in the field to see whether these assets that were once installed are still actually present.

Define Your Information Needs in Advance

It is very important that you define early on what information you truly need from these assets. If you try to collect too much information, such as every technical detail about a pump or perhaps the original purchase prices, it can take a lot of work to gather this and this phase of the project could take many months or even years. On the other hand, if you do not collect enough information, you may have to go back a second time to collect it later.

The most practical solution to this problem is to start with only the name and a few basic attributes of each asset, without even registering the components it consists of. You then use this as input for the criticality analysis, which you will learn about in the next chapter. Once an item proves to be critical, you can enter all the components and technical details for those specific assets.

Phased Approach

You can best work with a phased approach in which you gradually expand the information. In the first phase, you collect only the essential data: name, location, asset number, and primary function. This information is usually relatively straightforward to obtain and already gives you a workable foundation.

In the second phase, during the execution of maintenance work, you can collect additional information. Every time you carry out a work order on an asset, you can enter all the information you encounter into your enterprise asset management system. This way, you gradually build a complete database without slowing down your primary process.
It is very important that you work efficiently in this step. The engineer who performs the inventory should preferably also be the engineer who later conducts the analysis to set up the preventive maintenance program. This ensures continuity and prevents information from being lost between different people.

For TechPro's 200 assets, Sarah's inventory took 3.5 weeks. "I worked systematically," Sarah said. "First week: gathering all documents. Second week: field inspection per production line. Third week: entering and validating data. Half a week: resolving inconsistencies and prioritizing."

3.3. Implementing Quick Wins

While inventorying, you can already start working on quick wins by setting up basic inspection schedules that you know these assets will need regardless. One of the problems with setting up preventive maintenance in a structured way is that it takes time. Time that plant managers often don't want to wait for.

Therefore, you can give them quick results by implementing the basics of maintenance that you know will be necessary anyway. One of these basics could be periodic lubrication or a simple weekly inspection of the equipment. These types of tasks will have an immediate impact on the reliability of the equipment and the uptime of the assets. You do not need to wait until the entire project is finished before implementing them. At a later stage, you can still optimize these tasks.

3.4. Information Management and Coding

An important aspect to think about is how you are going to maintain all these records. Of course, you can enter a lot of information into the enterprise asset management system (EAM). You need to think about topics such as coding the equipment and structuring the database.

Systematic Numbering

The enterprise asset management system can simply start by assigning the first asset number 1, the second asset it finds number 2, and continue coding in this way so that every asset has its own unique number. Usually, however, it is easier to include some information in the code that people can also understand, such as giving every motor the letter M, every pump the letter P, and every valve the letter V, followed by some additional numbers that indicate which location or process the component belongs to, and then a unique number.

A good coding system helps not only with locating assets, but also with grouping them for maintenance planning. If all pumps start with P, you can easily select all pumps for a specific type of maintenance.

Consistency with Existing Systems

You have to be very careful not to invent numbers and put them in your EAM, while the process control systems and the P&IDs already have different numbers. It is easiest to match those numbers, because otherwise later, when you have a problem and an operator calls you and says there is a problem with pump 44 and in your EAM it is pump 128, confusion will arise.

This inconsistency leads to misunderstandings, incorrect work orders, and ultimately to reduced effectiveness of your maintenance program. Therefore, take the time to investigate which coding systems already exist in your organization and try to harmonize them.

Document Management

All of this information is not just about numbers and words. Much of it involves pictures, instructional videos, manuals, and PDF files that you may choose to scan and digitize if they are old documents. You can also choose to keep them on paper, but then you need to place them in a well-organized documentation system.

The trend is moving toward digitization, because this increases accessibility and reduces the risk of loss. Modern systems make it possible to link photos, videos, and documents directly to specific assets, which can significantly increase the efficiency of maintenance activities.

Figure 3-1 Technician inventories existing assets

3.5. Efficiency and Quick Wins

Developing this master equipment list can be one of the most time-consuming steps in the entire process of establishing preventive maintenance. But one thing to keep in mind throughout this entire process is: you have achieved nothing until the moment a technician or an operator actually works on the equipment and performs the maintenance. All the effort and time spent before that moment is necessary, and often it is better to do it all at once, but it produces no results until that moment when the first grease gun touches the first grease fitting.

Use of Technology

Modern technologies can significantly speed up the inventory process. Mobile apps make it possible to enter information on the spot, take photos, and scan QR codes. Some systems can even automatically detect assets via RFID tags or other identification technologies.

Consider also using external specialists for the initial inventory, especially if your organization is large or has complex installations. They have experience with efficient inventory techniques and can often work faster than internal employees who have to perform this task alongside their regular duties.

3.6. From Inventory to Prioritization

When your asset register is complete enough for the first phase, you can start determining which of all those assets are most important or even critical. You will learn this in the next chapter, where you will learn a quick but thorough method for ranking assets based on their impact on your operations.

The most important thing is that you now have a workable foundation for all further steps in your maintenance optimization. With a reasonably complete asset register, you can work systematically when developing your preventive maintenance program.

Sarah's inventory experience shows that this step, while time-consuming, is essential to the success of your entire preventive maintenance program. "That 3.5-week investment was the foundation for everything that followed," she reflected. "Without that thorough inventory, we never would have been able to identify our critical assets so effectively.

4. Critical Assets

Sarah gathered her team in the conference room: maintenance supervisor James, process engineer Lisa, production leader Mark, and operator Pete. "We're going to systematically assess all 200 assets," she began. "Three days of hard work, but then we'll finally know which machines are truly critical."

After the three days of 3-hour sessions each, they had a clear picture: 25 assets critical, 75 important, 100 standard. "Now the difference becomes clear," James said. "Not all assets are equal."
Now that you have defined clear requirements for your technical systems and have an overview of all your assets, the next step is determining which assets deserve the highest priority. Not all systems are equally important for your operations, and by categorizing assets based on their criticality, you can allocate your limited maintenance resources most effectively.

There are several methodologies for setting up or optimizing preventive maintenance for assets. Some of them are very time-consuming. The only methodology that is officially recognized by many standards and industries is Reliability Centered Maintenance (RCM). This is a very thorough methodology that is used, for example, to develop preventive maintenance schedules for new types of aircraft.

The problem with RCM is that it is very time-consuming. It is so time-consuming that some people call it a "Resource-Consuming Monster." In a typical organization, such as a production plant or a water treatment facility, you do not have enough time and budget for an RCM analysis for every asset.
That is why it is necessary to know how critical each piece of equipment is. For highly critical equipment, it may be necessary to perform a full classical RCM analysis, especially if this equipment is unique and you cannot rely on experiences from other industries. For less critical equipment, there are methodologies that are faster and require less manpower.

There is, however, an even more important reason to determine the criticality of each asset: not only do the more critical assets deserve more time, manpower, and money for the analysis of preventive maintenance, but also for the actual preventive maintenance itself. If equipment is highly critical and the consequences of failure are severe, it may be economically feasible to perform more maintenance tasks and to perform them more frequently than with less critical assets.

4.1. Criticality Ranking Methodology

Critical assets are systems whose failure has a major impact on essential business aspects. These aspects include production capacity, product quality, personnel or environmental safety, environmental impact, sustainability objectives, and regulatory compliance.
Which assets in your specific organization are critical depends on the requirements set by management that were discussed earlier. The risk matrix indicates how often certain failure events may occur.

Legal Requirements for Criticality Assessment

Various laws and regulations require organizations to know which assets are critical. For critical assets, you must perform a thorough analysis to identify possible failure modes and develop preventive measures. Companies must be able to demonstrate that they have performed this maintenance on time and correctly.

This legal obligation means that criticality assessment is not only a practical tool for the efficient allocation of resources, but also a requirement for compliance. Therefore, you must document how you determined the criticality and be able to demonstrate that your maintenance plan gives adequate attention to the most critical assets.

The Criticality Ranking Process

To determine the criticality of assets, you conduct a criticality ranking. This process takes place with a multidisciplinary team of three to seven people who have insight into how assets can fail and what the consequences are. A larger group often leads to less efficient discussions without substantial added value.

The process is based on systematically assessing all assets in your organization against a predefined set of criteria. These criteria stem directly from the requirements that management has established in the risk matrix. By assessing all assets against the same criteria, you obtain a consistent and objective ranking.

This scoring must be adapted to your specific organization and the amounts and timeframes that are relevant to your situation.

Performing a Criticality Assessment

To determine how critical each asset is, you need to be able to assess the consequences of that asset's failure against the requirements set by management. For that reason, you assemble a team that can assess the assets and their failure modes.

Team Composition

A typical team composition consists of a maintenance engineer with knowledge of asset design and maintenance techniques, a maintenance supervisor or experienced technician with practical knowledge, a process engineer who understands process requirements, a production manager with insight into operational impact, and an experienced operator who works with the equipment on a daily basis.

Valuable Perspectives from Each Team Member

Each team member brings valuable perspectives to the assessment. Mechanics and operators know the practical problems that occur and often have an intuitive sense of which assets are problematic. Engineers and production leaders know how often these problems occur and can estimate the frequency of failures. Production leaders and process engineers understand the impact on product quality and availability and know how failures propagate throughout the entire process.

The production supervisor knows whether workarounds are available when a specific asset fails, which is crucial for estimating the actual impact. Maintenance personnel can estimate the maximum duration of repairs, which is important for determining the total impact of a failure. Process engineers have insight into the impact on energy consumption and CO2 emissions, which is becoming increasingly important for sustainability objectives.

Sarah's team at TechPro was optimally composed. "James knew every machine inside and out," she said. "Lisa understood exactly how failures rippled through the process. Mark knew which machines we could afford to lose and which we couldn't. And Pete, our operator, had the best feel for which machines were truly problematic."

Conducting the Session

You conduct the criticality ranking session in a quiet room where the team systematically goes through the entire asset list. For each asset, the team determines the potential impact of a realistic worst-case scenario on the various aspects from the risk matrix.

Asset	Safety	Environment	Economic	Condition	Delivery Reliability	Priority
Heat exchanger HEX-58	3	2	5	4	5	Critical
Cooling water system	2	3	4	3	4	Critical
Main compressor	3	2	5	4	5	Critical
Packaging machine	1	1	3	4	3	Critical
Conveyor A12	2	1	2	2	2	Important
Hall lighting	2	1	1	1	1	Important
Office printer	1	1	1	1	1	Standard

Figure 4-1 Scoring table criticality ranking

A score of 1 in the safety column means that failure of the asset has no or minimal safety consequences, while a score of 5 indicates that failure can result in fatalities. At this stage, the team does not yet determine how often failures can occur — that depends on the maintenance plan still to be developed. The team also does not fill in all possible failure modes or causes. The only goal here is to identify assets whose failure consequences are serious. In many cases, teams also add a brief description of what these consequences are for scores of 4 or 5. The intention is not to perform extensive analyses of all assets. This is an initial screening to identify the critical assets that will need a thorough analysis later.

After completing the full list, a clear overview emerges of which assets are critical for various aspects such as economics, safety, quality, or environment. The matrix contains exactly those aspects that management considers important. Some organizations also include aspects such as delivery reliability, sustainability, or food safety.

Facilitation of the Process

You have the sessions facilitated by a facilitator. This must be someone who is independent from the other team members. He chairs the sessions and monitors the progress. The facilitator must know the methodology and ensures that the team stays focused on the task.
Sometimes people want to go too deep into all kinds of possible scenarios, or a discussion arises about a past incident. There is then a risk that the discussion turns into a blame debate, which gets in the way of an open conversation. In other cases, team members have insufficient information to be able to assign a well-founded score.
The facilitator must assess this, ask follow-up questions if necessary, and may decide that the team first needs to gather more information before filling in the score for a particular asset.

A session typically lasts two to three hours. If the sessions are shorter, members often feel that they have to stop just as they are hitting their stride. If the sessions run too long, however, attendees lose their concentration. An analysis requires intensive mental effort, and after three hours the quality of the decisions begins to decline.

4.2. Interpretation of Results

There are several methods for determining the criticality of assets, but for establishing preventive maintenance, the criticality ranking described here is the most effective. The method is sufficiently thorough without being unnecessarily time-consuming.
When setting up preventive maintenance, the primary goal of criticality ranking is to support the decision of which methodology you will use for the preventive maintenance plan for the asset.

Priority Categories

Based on the total scores, you classify assets into categories.
Assets that score 4 or 5 on one or more aspects are considered critical. In those cases, it is often worthwhile to have a team perform a full RCM analysis. There are also critical installations for which no RCM analysis is necessary. In many cases, for example, a steam boiler within a company may be critical, yet not require an RCM analysis. The hazards associated with a steam boiler are significant, but that is precisely why various experts and legislators have already conducted analyses and established maintenance requirements. The company must assess for itself whether its steam boiler is "standard" enough to make an RCM analysis unnecessary. The point here is that the score is a tool, not an automatic trigger.

Assets that score 2 or 3 on one or more aspects are considered important. A reliability engineer analyzes these assets independently, not as part of a team. He does discuss his results as a team, however, and checks whether they are correct and accepted by his maintenance and production colleagues. This approach forms an important element of Industrial RCM (iRCM).

Assets that score 1 on all aspects are considered standard. These assets receive basic maintenance according to standard procedures that the reliability engineer sets up. In doing so, he consults experts and reads manuals, but group sessions are not necessary. In many cases, he discovers that the maintenance prescribed in maintenance manuals is more extensive than necessary.
Sarah's team methodically scored TechPro's 200 assets. "The main compressor received a score of 5 economically — failure would cost $3,500 per hour," Sarah explained. "The office printer scored 1 across the board. The difference was immediately clear: 25 critical, 75 important, 100 standard."

Other Applications of Criticality Ranking Results

When setting up preventive maintenance, you use the results of the criticality ranking as described above.
Companies, however, use the results for even more purposes. In new construction projects, the criticality ranking can already take place when the very first basic design of a process or production line is ready. At that point you know what kind of machines or equipment are needed, but not yet their design. For critical applications, it is worthwhile for the company's engineers to be involved in the design process. It is dangerous to purchase a standard off-the-shelf asset without thoroughly analyzing it. For less critical assets, it is sufficient to simply purchase an asset with good quality.

During the execution of maintenance, you can use the results of the ranking to determine which maintenance gets priority. Maintenance on less critical assets may sometimes be postponed, maintenance on critical assets may not.
A third application is in setting up or auditing any kind of management system. The ranking makes clear which assets are critical for, for example, the environment. This is one of the first things an environmental auditor will ask about.

Preparation for the Next Phase

In the next phase, you will analyze the most critical assets in much greater depth. The primary goal of this initial analysis is to prioritize further maintenance development. In practice, it is usually not feasible to analyze every asset with equal thoroughness.
By prioritizing this way, you can focus your limited time and resources on the assets that make the greatest difference to your operations. This leads to a far more effective maintenance plan than if you were to treat all assets equally.

Efficiency of the Methodology

A significant advantage of this criticality ranking method is speed. A typical session lasts approximately three hours, during which a team can assess several hundred assets. Alternative methodologies such as Failure Mode and Effect Analysis (FMEA) take considerably more time. With FMEA, a team may need a year just to determine which assets are most critical before they can even begin developing maintenance strategies.

During this analysis phase, you perform no concrete maintenance actions, which means that potential improvements can only be realized much later. The criticality ranking offers a pragmatic balance between thoroughness and efficiency.

This speed also means you can repeat the assessment regularly. If your business conditions change, you install new assets, or you gain new insights about the impact of failures, you can update the criticality ranking without a major time investment.

Quality Assurance of the Process

To ensure the quality of your criticality assessment, it is important to document the process thoroughly. Record which criteria you used, who was on the team, and what information the scores were based on. This helps with later reviews and makes it possible to update the assessment when circumstances change.

It is also valuable to validate the results by checking whether the outcomes align with your intuitive sense of which assets are critical. If the methodology flags assets as critical that no one expected, or overlooks assets that everyone considers critical, it is worthwhile to review the criteria or the scoring.

4.3. From Prioritization to Method Selection

Now that you know which assets are critical and which have lower priority, in the next chapter you can determine how to develop the optimal preventive maintenance for each of these categories. Different criticality levels require different approaches, ranging from extensive team analyses for the most critical assets to efficient standard procedures for less critical systems.

The criticality ranking you have now completed forms the basis for all further decisions about your maintenance plan. It enables you to allocate your resources effectively and ensures that you use the right method for each asset, depending on its criticality to your operations.

5. Determining Preventive Maintenance

Sarah stared at the list of 25 critical assets. "We'll start with the most important one," she decided. "Centrifugal pump P-401 — the one that costs $3,500 per hour when it fails." This would be Sarah's first complete RCM analysis. Over 2 sessions of 3 hours each, she would work together with her team to develop a thorough maintenance plan for TechPro's most critical asset.

Now that in chapter 4 you have established which assets are critical to your organization, you can determine the optimal maintenance strategy for each asset.

5.1. Maintenance Methodologies - Overview

For determining optimal preventive maintenance, you use three methodologies, each with their own area of application. The choice between these methodologies depends on the criticality of your assets and the available resources in your organization.

Reliability Centered Maintenance (RCM) is applied to highly critical installations where failure would have serious consequences. RCM is carried out with a full team of people. It is time-intensive but extremely thorough. In addition, RCM is used for assets of which an organization has a large number — a thorough analysis that may lead to only a small cost saving **per asset will still pay for itself in that case.**

Industrial RCM (iRCM) is a streamlined version that takes less time. You apply this to important, but not critically vital assets. The methodology retains the core principles of RCM, but simplifies the process. An independent engineer does the preparatory work, after which a team further develops a number of points.

The independent approach is used for other, less critical assets. As a reliability engineer, you develop the maintenance program independently, with informal consultation where needed. This approach is the least time-intensive but still systematic.

The choice between these methodologies determines not only the time investment, but also the depth of the analysis and ultimately the effectiveness of your maintenance plan. For critical assets, the additional time investment of RCM is justified by the potential savings and risk reduction. In the following section, you will first read about RCM. Once you understand this comprehensive approach, you will also understand the logic behind the other two methodologies.

5.2. Reliability Centered Maintenance (RCM)

What is RCM?

Reliability Centered Maintenance is a systematic methodology for developing preventive maintenance schedules. The end result is a detailed schedule that specifies what maintenance a technical system requires.
Originally developed for commercial aviation, RCM has been adapted for various industries. General industry typically uses simplified variants, but the core principles remain the same.
In an RCM analysis, you systematically map out with a team of experts what requirements the organization places on an asset, how the asset can fail to meet these requirements, what the causes of failure are, and what preventive measures are possible.

Team Composition for RCM

An industrial RCM team typically consists of approximately five members. As a maintenance engineer, you bring knowledge of asset design and maintenance options. A breakdown technician has hands-on experience with the asset. He knows exactly what happens when a failure occurs. A user or operator knows what happens when the asset fails, including early warning signals. A production manager has insight into process impact and can assess the consequences of a breakdown. A facilitator guides the process and ensures structure.

The facilitator has multiple tasks: chairing the team, monitoring the correct application of the RCM methodology, promoting efficient working practices, and ensuring proper registration of results.
The facilitator does not need to know the asset or system personally — unfamiliarity can even be an advantage, as it leads to clarifying questions.

However, the facilitator must be independent and have some general technical knowledge in order to follow and guide the discussions.

The RCM Methodology: 7 Core Questions

RCM follows a fixed framework in which you systematically answer seven core questions. These questions form the backbone of every RCM analysis and ensure consistency and completeness.

1. What are the functions and performance requirements for the technical system in its current application?
2. In what ways can the system fail to fulfill these functions?
3. What causes can lead to these functional failures?
4. What are the effects of each failure mode and what exactly happens?
5. What are the consequences of failure and how severe is the impact?
6. What preventive measures can we take to prevent failure or detect it early?
7. What should we do if we cannot find a suitable preventive measure?

RCM distinguishes between different maintenance types.

Failure-based maintenance (FBM) follows the "run-to-failure" strategy.

Interval-based maintenance (IBM) means periodic replacement based on a fixed interval, such as time, operating hours, miles driven, or number of products produced.
Condition-based maintenance (CBM) is inspection to assess condition, followed by maintenance only when needed.
Function tests are periodic tests to verify whether components are still functioning correctly.

Performing an RCM Analysis

You carry out an RCM analysis in a quiet environment where the team can work without interruption.The optimal session length is approximately three hours — longer leads to fatigue and reduced decision quality.

Between sessions, there should be a maximum of one week to maintain continuity.
The room must be equipped with visualization capabilities such as a monitor for RCM forms and ideally a second screen for technical information such as P&IDs and technical drawings. The process begins with a clear definition of scope. For complex systems, the facilitator divides the system into logical sections that are analyzed one at a time. This prevents the team from being overwhelmed by the complexity.

Complete RCM Case Study: Centrifugal Pump Analysis

To illustrate the RCM methodology, we follow Sarah's team through their complete analysis of centrifugal pump P-401 in the cooling water system. This pump is critical to the production process, because failure leads to an immediate production shutdown.

Situation Overview and Team Composition

The centrifugal pump P-401 has a replacement value of $15,000 and pumps 40 m^3/h of cooling water at 2.5 bar for a production line. When it fails, the entire line must be shut down at a cost of $3,500 per hour, plus possibly a $12,000 rejected batch. The pump is eight years old and has recently been showing increasing vibrations.
Sarah's RCM team consists of herself as maintenance engineer, Rob (experienced breakdown technician), Maria (process engineer), James (production manager), and Tom (external facilitator).They gather in the conference room with a large screen on which the RCM form is projected.

Determining the Scope of the Analysis

Before moving on to the first RCM question, Tom picked up a marker and drew a simple diagram on the whiteboard. "First, we need to determine exactly what we're going to analyze," Tom said. "What falls within our scope and what doesn't?"
James looked at the diagram. "Do you mean we don't have to look at the entire installation?"

"That's right," Tom said, nodding. "RCM is most effective when you focus on a clearly defined system. Too broad and you get lost in the details, too narrow and you miss important connections."
Sarah drew the pump with its main components on the board. "I suggest we take the pump itself, including the motor, coupling, bearing housings, and the direct pipe connections."

Rob pointed to the diagram. "And what about the variable frequency drive? Isn't that directly connected to it?"
Maria shook her head. "That control system is actually a separate system. If we include it, our analysis will become far too large. I suggest we address it later in a separate RCM session."
Tom drew a clear line around the pump and its direct components. "Agreed. So our scope is pump P-401, including the motor, coupling, mechanical components, and the connecting flanges. We'll leave the electrical controls and the further piping for another time."
The team nodded in agreement. The scope has been clearly defined and they can move on to the first real RCM question.

Question 1: Functional Requirements and Performance Standards

Tom said, "Let's start with the first RCM question. What exactly are the functions of pump P-401?"
Maria said, "The primary function is pumping 40 m^3/h of cooling water at 2.5 bar pressure. This cooling water is essential for our extruder."
Rob said, "Sorry Maria, can you explain what you mean by primary function? And are there secondary functions as well?"
"Good question, Rob. Primary functions are the main functions for which we acquired the asset — so the reason the pump is there. Secondary functions are other functions that the pump must fulfill, such as containing the fluid, preventing leaks, and providing process information. Requirements are conditions that the asset must meet in order to function properly. Think of safety requirements, environmental requirements, or constraints such as noise level limitations."

"Ah, now I get it," Rob said. "So the primary function is why we have the pump, and a secondary function could be that the pump must not leak? It can therefore happen that the secondary functions are more important than the primary ones, for example when it involves a pump that handles hazardous substances?"

"Exactly," Tom said. "Let's continue."

Sarah said, "We also need to define the secondary functions. The pump must show no leakage, must remain within the noise standards of 80 dB, and must not consume more than 15 kW."

Rob said, "Don't forget that the pump needs to be explosion-proof — we're in an ATEX Zone 2 area."

The team systematically worked through all the functions. They entered the functions in the first column of a spreadsheet. Specialized software exists for this purpose as well, but they deliberately chose to start simple.

Functions
Pumping liquid at 38–42 m³/hr and 2.3–2.7 bar
Must not leak
Noise < 80 dB(A)
Energy consumption max 15 kW
Bearing housing temperature max 60°C

Figure 5-1 Function table

For the primary function, they establish that the pump must move material at a flow rate of 38-42 m^3/h at a head pressure of 2.3-2.7 bar. For secondary functions, they define that there must be no visible leakage, the noise level must remain below 80 dB, the energy consumption may not exceed 15 kW, and the temperature of the bearing housings must remain below 60°C.

Question 2: Identification of Functional Failures

Tom led the team to the second question. "Now that we know what the pump is supposed to do, how can it fail to perform these functions?"

"You mean we need to list possible causes of failures now?" Rob asked.

"That's not really what I mean," Tom explained. "We need to make a list of possible failures that can occur. You don't actually need technical knowledge for this. Logical reasoning is sufficient. If the pump needs to move fluid within a certain range, then it's a failure if it pumps too little."

"Yes, that makes sense," Maria laughed.

"But it's also a failure if it pumps too much. That can have very different causes. It is important to do this step carefully, otherwise we might overlook possible causes in the next steps."

The team systematically identified all possible functional failures and entered them in the next column of their spreadsheet.

Functions	Functional failure
Pumping liquid at 38–42 m³/hr and 2.3–2.7 bar	Pump does not pump
	Pump delivers less than 38 m³/hr
	Pump delivers more than 42 m³/hr
	Pump delivers less than 2.3 bar pressure
	Pump delivers more than 2.7 bar pressure
Must not leak	Pump leaks
Noise < 80 dB(A)	Noise > 80 dB(A)
Energy consumption max 15 kW	Energy consumption > 15 kW
Bearing housing temperature max 60°C	T>60°C

Figure 5-2 Table of functional failures

Question 3: Causes of Functional Failures

This will be the most time-consuming phase of the analysis. The team must systematically identify all possible causes for each functional failure. For "pump does not pump or pumps insufficiently," the team uses a top-down approach. Sarah drew on the whiteboard. "A centrifugal pump works because an electric motor drives an impeller that displaces fluid through rotation."

"So if the pump isn't pumping, then either the motor isn't running, or the impeller isn't turning, or the impeller can't move fluid," Rob said.

The team works this out systematically: the motor does not run due to no electrical power, motor windings defective, rotor blocked, output shaft broken, or motor bearings seized. The drive fails due to coupling becoming loose, drive shaft broken, or pump housing mechanically blocked. The impeller cannot move fluid due to impeller damaged or come loose, cavitation due to insufficient inlet pressure, air in the system. The causes they found, they entered in the following columns of their spreadsheet. They discovered that it is very important to work systematically. In the beginning, they regularly found causes that they could not properly place in their sheet, or that belonged in a branch they thought they had already worked out, but which still turned out to be incomplete.

For each main cause, they continue working down until they reach a level where concrete maintenance actions are possible. For "Pump bearings seized," they continue: this can be caused by insufficient lubrication, overloading, or incorrect installation.

For "insufficient lubrication" they investigate further: grease nipple clogged, normal grease consumption, wrong lubricant used, and a number of additional causes.

"We need to stop at a level where concrete actions are possible. With 'normal grease consumption', we can determine how often lubrication is normally needed."

After two hours of intensive work, the team had identified 47 different failure modes for all the functional failures of the pump.

Cause level 1	Cause level 2	Cause level 3	Cause level 4	Cause level 5
Impeller does not rotate	Impeller blocked	Build-up inside pump	CIP problems	Analyzed in the RCM of the CIP
		Foreign object	Does not occur	
		Pump bearings seized	Surface fatigue	
			Moisture in bearings	Defective seals
			Lack of lubrication	Grease nipple blocked
				Normal grease consumption
				Wrong lubricant
			Overload	Misalignment
				Cavitation
				Imbalance
			Incorrect assembly	
	Coupling came loose	Has never occurred		
	Drive shaft broken	Has never occurred		
	Impeller came loose from shaft	Has never occurred		
	Electric motor does not run	No power supply	Analyzed in RCM of power supply	
		Windings defective	Normal aging	
			Motor overheating	Cooling fan not running
				Ambient temperature too high
				Motor overloaded
				Cooling fins contaminated with dust
		Rotor blocked		
		Bearings seized	Surface fatigue	
			Moisture in bearings	Defective seals
			Lack of lubrication	Grease nipple blocked
				Normal grease consumption
				Wrong lubricant
				Grease pump defective
			Overload	Misalignment
				Cavitation
				Imbalance
			Incorrect assembly	
Pipes blocked	Product build-up	CIP problems	Analyzed in the RCM of the CIP	
Impeller cannot displace liquid	Cavitation	Inlet pressure too low		
	Air in the system			
	Impeller damaged			

Figure 5-3 RCM cause form

Question 4: Effects of the Failure Modes

For each failure mode, the team describes exactly what happens and how it can be observed.

For "motor windings defective due to normal aging," Rob said, "Let me try to walk through it in order. First, we can detect the degradation by measuring the insulation resistance across the windings — we call that meggering. As the aging progresses, the temperature increases due to

elevated resistance. We can observe that with an infrared measurement. If we then fail to intervene, the motor eventually gets so hot that the windings deteriorate faster and faster and ultimately short out."
The team records for each failure mode the time elapsed between the first detectable signals and the eventual failure. This P-F interval is crucial for determining inspection frequencies. You can read more about the P-F interval and determining the optimal maintenance frequency in chapter 8.
“How large the P-F interval is can often be difficult to say,” Tom said, pointing to the next section of the form. “It determines, however, how often we need to perform these maintenance activities.”
Rob frowned. "What do you mean? Isn't that just in the supplier's manual?"

"For some things, yes," Sarah replied. "With standard condition monitoring we have fixed intervals - we measure vibrations monthly or quarterly, for example, and we do oil analysis every six months."
"But for many other failure modes, it's not that clear," Maria added. "Then we have to estimate ourselves how quickly a problem develops from just detectable to an actual failure."
"Exactly," Sarah said. "With RCM, we can determine particularly well which maintenance actions we need to perform, and somewhat less precisely how often. That also depends each time on the conditions under which a system operates, on the make and type, on how heavily it's loaded, and so on."

Rob nodded. "You can't put a fixed interval on that, indeed."
James sighed. "And that's exactly what our production manager always asks. He doesn't want to know what maintenance we're going to perform, but how often we need to do it. He just doesn't realize how significantly the content of the maintenance actions shifts as a result of this analysis."
Tom picked up his marker: "We are now going to go through each maintenance activity and discuss which scenarios can occur in your specific situation. Some intervals we can base on experience, others on manufacturer recommendations, and still others we will need to estimate and adjust later."

"It really is tricky," Maria admitted, "but also incredibly important. Too much maintenance costs unnecessary money, too little maintenance increases the risk of failures."

Question 5: Consequences of the Failure Modes

James took the lead in estimating the consequences. "If this pump fails, we'll have to shut down the entire production line. That costs us $3,500 per hour in production losses plus ongoing costs."
Maria added, "If we don't act quickly, the extruder could overheat. Then we'd have to reject the entire batch — that's an additional $12,000."
The team systematically categorizes all consequences by economic impact, safety consequences, environmental impact, and quality consequences, using the risk matrix.

Question 6: Selection of Preventive Measures

Now the team arrives at the core of RCM: what preventive measures are possible and effective? This question forms the heart of the RCM methodology. RCM uses a structured decision diagram that guides you step by step through all possible options. This diagram begins with a crucial first question that determines which path you will follow: "Are the consequences of this failure hidden?"
A hidden failure is a failure that, when it occurs on its own, has no consequences for the functioning of the asset. Only when a second failure occurs does the function of the entire system fail. The classic example is safety systems:

- A broken emergency stop button causes no problems... until you need it
- A sprinkler system that fails has no consequences... until a fire breaks out
- A backup system that doesn't work is not a problem... until the main system fails

An evident failure, on the other hand, has noticeable consequences when it occurs, sooner or later.

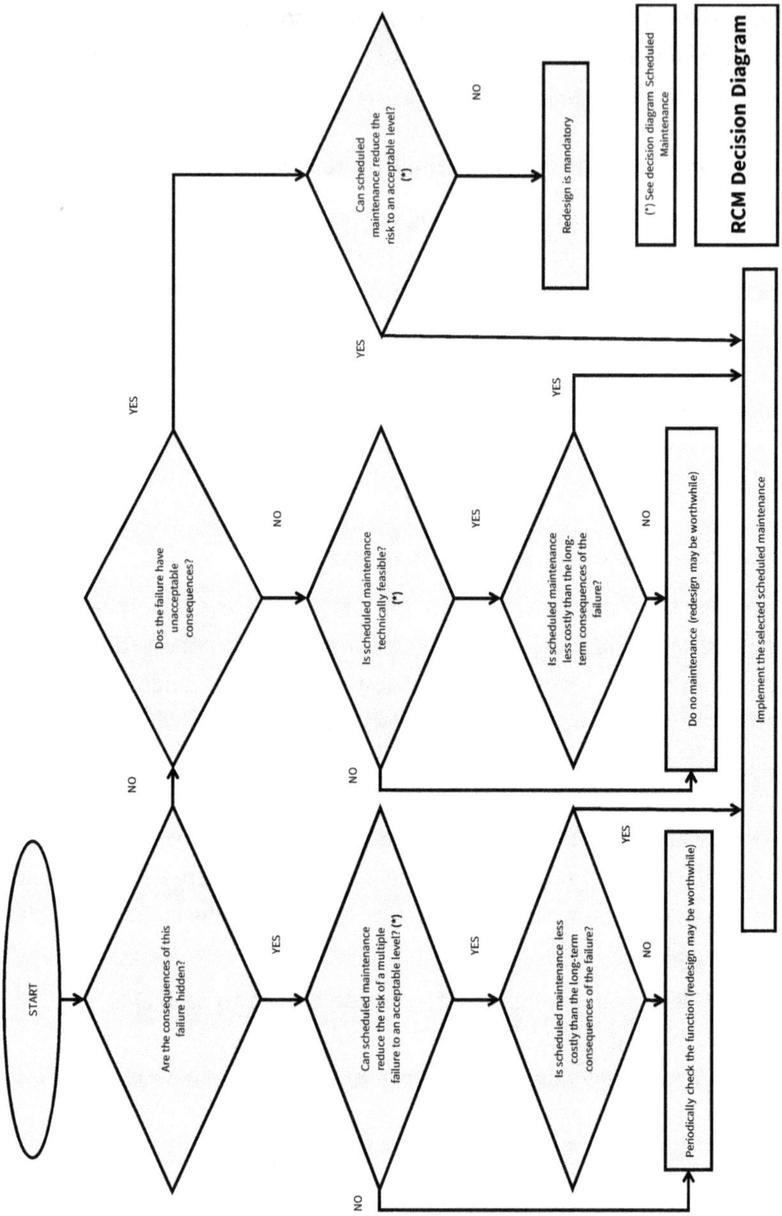

Figure 5-4 RCM Main decision diagram

For evident failures, RCM asks the following question: "Does this failure have unacceptable consequences?"
You determine what is unacceptable based on the risk matrix that was previously established in your maintenance policy. Unacceptable consequences are usually related to safety or environmental impact.

This dichotomy leads to three main paths in the decision diagram:

1. Hidden Failures
2. Evident and acceptable
3. Evident and Unacceptable

Each of these three paths leads to a diamond that poses a slightly differently worded question. This question refers to decision diagram two, which in practice should actually appear in each of the three paths on the main diagram.

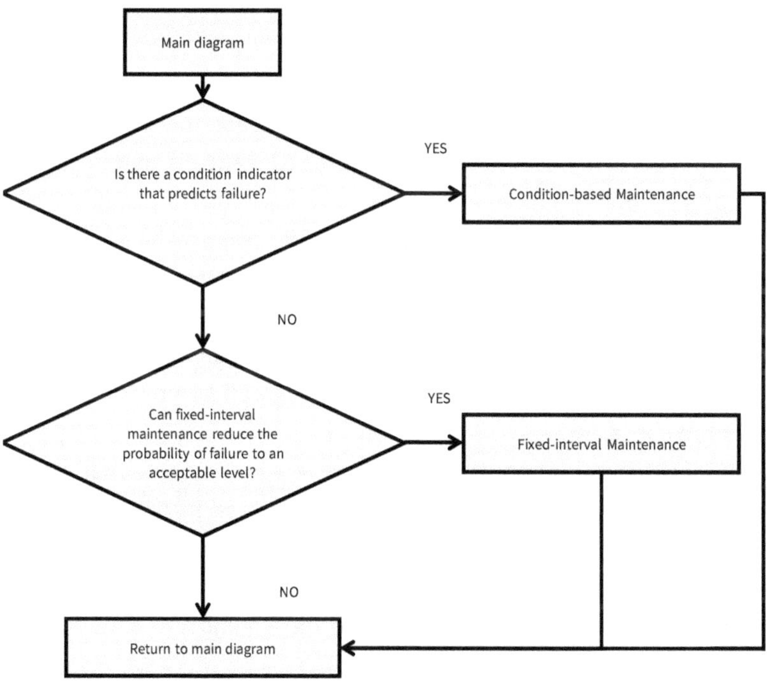

Figure 5 5 RCM decision diagram scheduled maintenance

This diagram poses two questions.

Question 1: **Does this failure mode have a condition indicator that predicts failure?**

This means: can we determine through measurement, inspection, or monitoring that the failure is developing, before it actually occurs?
occurs?
RCM deliberately favors condition-based maintenance for several reasons:
Optimal timing: You perform maintenance exactly when needed, not too early and not too late. With usage-based maintenance, you always have to be 'safely early' and often waste useful remaining service life.
Lower costs: An inspection is usually less expensive than the preventive replacement of a part that might last for years to come.
Fewer consequential damages: every time you work on an asset, you increase the risk of damage, for example through assembly errors.

An inspection is far less invasive than replacing or overhauling part of an asset. The more invasive the maintenance, the greater the chance of problems afterward. A well-known saying in maintenance is therefore: "If it ain't broken, don't fix it!"
Extra benefits: During inspections, you often discover small problems that are easy to fix, but if left untreated can lead to costly failures. Think of loose bolts, poor alignment, or early-stage damage to other components.
Actual knowledge: You learn which failure modes truly occur in your specific circumstances, instead of working with theoretical intervals that may be far too conservative.
For the centrifugal pump, the team discovers for example that "motor bearings seized due to insufficient lubrication" is highly predictable: monthly vibration measurements can detect a number of bearing problems 2 to 3 months in advance. How far in advance exactly is highly dependent on the specific cause of failure and a number of other factors.

Question 2: **"Can maintenance after a fixed interval reduce the probability of failure to an acceptable level?"**

Only when condition-based maintenance is not possible does RCM consider the second option: usage-based maintenance.
This means replacing parts based on a fixed interval (time, operating hours, miles, production cycles) regardless of their actual condition.
Sarah's team considers, for example: "We could preventively replace the bearings every five years, regardless of their condition." But because vibration measurements are possible, they prefer condition-based maintenance.

Back to the Main Path: The Follow-up Decisions

From the nested diagram, you return to your main path with one of three conclusions:

1. Condition-based maintenance is possible
2. Usage-based maintenance is possible
3. Neither maintenance type is practically feasible

For unacceptable evident failures: Redesign mandatory

If you are on the "evident and unacceptable" path and no maintenance is possible, the conclusion is clear: redesign is mandatory.
In other words, an unacceptable failure can occur that you cannot prevent with maintenance. Then you must modify the design or the operating method of the asset. Safety, environmental, and other critical aspects take precedence over economic considerations.
If preventive actions are possible, you must carry them out. In doing so, you do not make an economic trade-off between the cost of maintenance and the cost of a failure. After all, the latter is unacceptable. Of course, it may still be economically worthwhile to modify the design.

For Acceptable Failures: The Economic Trade-off

For hidden failures and acceptable evident failures, RCM asks one more important question: "Is standard maintenance less expensive in the long run than the consequences of the failure?"
Pay attention to the phrase "over time" — it's not about the cost of a single failure versus a single maintenance action, but about the cost per year or per period.

For acceptable evident failures: If maintenance is more expensive than the damage caused by the failure, RCM advises: "Do no maintenance" or "run to failure." Let the failure occur and then repair it.

Rob struggled with this at first. "But then we're deliberately letting things break down!" Tom explained. "That's right, but only when it's economically sensible and the consequences are acceptable. For a light bulb in an office, that makes sense — replacing it preventively costs more than the inconvenience of failure."

For Hidden Failures: Function Tests as an Alternative

With hidden failures, you have a unique opportunity: function tests. This means periodically testing whether the failure mode has already occurred.

This differs from condition-based maintenance: with function tests you check whether something has already failed, whereas with condition-based maintenance you inspect whether something is going to fail in the future.

Examples of funtion tests:

- Monthly testing of emergency stop buttons
- Weekly inspection of smoke detectors
- Periodic testing of backup systems
- Testing the operation of safety instrumentation
- Periodic startup of the emergency power installation

Question 7: When No Preventive Measure Is Possible

For some failure modes, the team concludes that no economically justifiable preventive measure is possible.

Rob said, "This happens so rarely that prevention is more expensive than occasionally replacing a motor."

Tom added, "But we also need to look at the consequences. If a lightning strike only damages the motor, we can opt for failure-based maintenance."

Sarah countered, "But what if a lightning strike also damages our process controls? Then we'd suddenly have much greater damage."

"Then we need to modify the system," Tom replied. "If the consequences are unacceptable and maintenance doesn't help, then re-engineering is necessary. We could, for example, install surge protection."

The Result

The result of the team's analysis is an overview of all functions, possible failures, causes, effects, consequences, and possible maintenance actions for each asset. It works best to keep the columns together in a single sheet. It is not possible to print such a sheet in a book like this. In Figure 5-6, you can see what the team entered after the cause 'aging of the windings of the electric motor.' The team performed this analysis for every possible cause. It may seem very time-consuming, but in practice it takes only a few hours. This costs less than the impact of a single failure.

RCM effects and consequences for aging of electric motor windings

Failure effect	Failure consequence	E/V	E/M	HSC	Int	n TAO GAO SAO	Maintenance description	Interval (in weeks)
Insulation resistance decreases (detectable by megger testing during shutdowns). Temperature gradually rises (measurable with IR during operation). With continued degradation: further heating leads to short circuit and motor shutdown.	3 hours downtime = $24,750 loss	E	No	Yes	No	TAO	Thermographic inspection of motor	4
						TAO	Megger testing of windings	52

Figure 5-6 RCM Effects and consequences

Of course, you still need to develop each action into a work instruction before you can execute it. You do that in a later step. It would take too much time to develop every maintenance task in detail. For many tasks it is also not necessary, and for the technicians performing them a title alone is sufficient. You can read more about this in chapter 9.2. It is also necessary, during the group session, to already record whether a production shutdown is required for this maintenance task and who will perform it. By that I don't mean which specific person, but which department or which company. You need this information to be able to cluster the maintenance tasks into efficient work orders in a subsequent step.

The Value of Redesign

Redesign or re-engineering is often an excellent way to prevent failures. That is why it is ideal to conduct an RCM study during the design phase of

an asset. At that point, you can still consider various design options and potentially avoid very costly maintenance.
Also with existing assets you can of course choose redesign, but this does carry risks. RCM was originally developed for aviation, where modifications to existing aircraft take a long time and cost a lot of money. Moreover, there is the risk that you solve one problem, only to end up with another one in its place.
For the pump case, the team discusses both options: for failure modes with acceptable consequences, they choose condition-based maintenance, supplemented by rapid repair procedures and stock of critical spare parts. For potential failure modes with unacceptable consequences, they consider system modifications such as redundancy or additional safeguards.

The Logic Behind RCM Decisions

The RCM methodology ensures that you always make the most logical choice:

Safety first: Unacceptable consequences always require action, regardless of cost. When RCM was developed for aviation, unacceptable consequences referred only to consequences for the safety of aircraft. Later, unacceptable consequences for the environment were added. Today, it encompasses all unacceptable consequences, such as those related to sustainability, quality, reputation, et cetera.

Optimal Maintenance Strategy: Condition-Based Is Preferred Over Usage-based

Economic Trade-off: Only When Consequences Are Acceptable Do We Weigh Costs Against Benefits

Realistic alternative: "Do nothing" is a legitimate choice when it is economically sound

This logic explains why RCM is so effective: it prevents both over-maintenance (too much prevention) and under-maintenance (too little prevention) by systematically supporting every decision.

Results of the Analysis

After two sessions of three hours each, Sarah's team developed a complete maintenance plan for pump P-401. The plan includes:

- Monthly Vibration Measurements
- Monthly Thermography
- Annual insulation measurements
- Bearing lubrication every six months per schedule
- Annual inspection of coupling and alignment

James was enthusiastic. "If we follow this plan, I expect we can increase reliability by 40% and at the same time reduce maintenance costs by 25% through fewer unplanned failures."
"The RCM analysis was intensive but valuable," Sarah reflected. "From uncertain in the first session, I grew increasingly confident in the methodology. Now I understand why P-401 deserves this thorough maintenance plan."

Lessons from the Case Study

This case study illustrates both the power and the time investment of RCM. For a single pump, six hours of team time were required, but the result is a thoroughly substantiated maintenance plan tailored to the specific criticality and circumstances.
For highly critical assets, this investment is justified. For less critical assets, more efficient approaches are necessary, such as Industrial RCM.

5.3. Industrial RCM (iRCM)

For organizations with hundreds of similar assets, a full RCM analysis per asset is not practical. Industrial RCM offers a more efficient approach that preserves the core principles while requiring less time.

iRCM Approach

Industrial RCM (iRCM) was developed to overcome the limitations of classical RCM analyses. Classical RCM is time-consuming, because each team analyzes every installation separately, which can lead to inconsistent

results for comparable assets. In addition, classical RCM offers few concrete tools for re-engineering when preventive maintenance is not effective.
With iRCM, as an engineer you perform one thorough analysis for a specific type of asset. This approach is based on the insight that similar components largely share the same failure modes and maintenance needs. Two centrifugal pumps of the same type will, for example, have virtually identical failure modes, regardless of which installation they are in.

Step 1: Creating a complete template

The foundation of iRCM consists of building a complete template for each component type. For every standard component, a full RCM analysis is performed once, in which all possible failure modes and associated maintenance actions are identified. This template contains all maintenance actions that can technically contribute to maximizing reliability.

An example of such a complete template for gear motors could consist of checks for soft foot during vibration measurements, visual inspection for soft foot, checking whether the fan is blowing, fan inspection and cleaning, periodic gauging where applicable, periodic lubrication where needed, infrared measurements, checking for thermal blockages, motor current analysis, oil changes, periodic oil sampling, vibration measurements on bearings in the speed range, and ultrasonic inspection of bearings. This list therefore truly encompasses all conceivable maintenance actions that can improve the reliability of a geared motor.

Step 2: Sorting Based on Criticality Scores

Next, you retrieve the scores from the criticality ranking you previously performed for your installation. You sort all assets of the same type based on their score. A gearmotor in a critical main pump, for example, receives the same score as that main pump. This systematic sorting forms the basis for developing different maintenance concepts.
The categorization varies by company, but a typical distribution might be that all assets scoring 4 or 5 are considered critical, assets scoring 2 or 3 are considered important, and the rest are considered standard. You

determine these thresholds based on your organization's specific risk tolerance and resources.

Step 3: Shortening the Template

Now you determine what maintenance you will provide for all assets in each category by systematically scaling down the complete template. For critical gear motors you will likely perform almost all actions from the template: monthly vibration measurements, quarterly visual inspections, annual infrared measurements, three-yearly oil changes with laboratory analysis, and so on. For important gear motors you scale this back to, for example, quarterly vibration measurements, semi-annual visual inspections, and three-yearly oil changes without laboratory analysis. For standard gear motors you might choose only annual visual inspections and five-yearly oil changes, or even fully run-to-failure maintenance.
This process of "cut and paste" is in principle so systematic that in theory a computer could carry it out. There are now software tools on the market that help with assigning maintenance concepts to assets based on their criticality.

Step 4: Assignment and Exception Control

In practice, however, it is always necessary to check on a per-asset basis whether there are any special circumstances, conditions, or components that require an exception to be made. A geared motor operating in an extremely dusty environment, for example, requires more frequent cleaning, while a unit in a clean room can operate with less maintenance. The advantage of this systematic approach is that you have a well-founded maintenance plan for each asset without extensive calculations per individual asset. You combine the thorough technical analysis of the template with the practical prioritization based on criticality.

Team Validation and Refinement

Next, you discuss this preparatory analysis with a multidisciplinary team. The team improves and supplements the prepared analysis, performs analyses for components where multidisciplinary input is needed, and discusses which maintenance actions are meaningful for the organization.

The team validation focuses on crucial questions such as whether all proposed actions are practically feasible under the specific circumstances, whether specific checks are missing that are relevant to this application, whether unique components exist that require custom analysis, and whether the proposed maintenance is economically justified. Also considered are historical failure patterns and possible system improvements. It is always wise to also visit the site during the analyses, for example to check the accessibility of the asset.

Although not all individual assets receive an extensive iRCM session, the team can during these sessions make a decision about the maintenance regime for all motor reducers in the company. This makes the approach highly efficient.

For situations where preventive maintenance is not effective, the team discusses alternatives such as run-to-failure maintenance with rapid repair procedures or system modifications to reduce the likelihood of failure.

This approach reduces team time to approximately six hours per asset type, because the preparatory analysis is done individually. The team focuses on validation, improvement, and decision-making rather than building the analysis from scratch.

5.4. Independent Approach

For routine assets, as an experienced reliability engineer, you can develop the maintenance independently, with informal consultation where needed. This saves time compared to formal team sessions.

The independent approach has two disadvantages: potentially less in-depth analysis due to more limited input, and lower acceptance due to less stakeholder involvement.

To create sufficient buy-in, you organize several team sessions so that stakeholders gain insight into the methodology. In practice, a reliability engineer develops approximately 90% of the maintenance plan independently, while 10% is worked out in a team setting.

This approach is suitable for assets with low criticality where standard maintenance methods suffice and where the risks of suboptimal choices are limited.

5.5. Method Selection per Asset Category

Now that you are familiar with all three methodologies, you can make a deliberate choice. For assets with, for example, a score of 5 on one or more criteria, you use full RCM (12 hours of team time). For assets with scores of 4, you use Industrial RCM (6 hours of team time). For assets with a maximum score of 3, you use the self-directed approach.
This differentiation ensures that you optimize your limited time and resources. You invest the most time in the assets that make the greatest difference for your organization, while still treating all assets systematically.

Companies that have less time available often choose to first complete all RCM analyses. These are necessary to prevent calamities. After that, they schedule, for example, ten sessions per year, to gradually analyze all their important assets. The disadvantage of this approach is that some important assets may wait years before receiving the analysis they need, and that during all that time the companies are wasting money and time on failures they could have prevented.
In chapter 6 you explore the fundamental causes of failures and failure behavior, which is essential for correctly applying the maintenance methodologies from this chapter.

6. Causes of Failures and Failure Behavior

Sarah was faced with a puzzle. TechPro's geared motors were failing far more often than expected — some after two years, while others ran without problems after eight years. "Why is there such a difference?" she wondered. By systematically analyzing the failure causes, she discovered that 60% of the failures had human causes: incorrect lubrication, vibration due to poor alignment, or overloading during startup. "We can dramatically improve this service life," she concluded.

Now that you have learned in chapter 5 how to systematically develop preventive maintenance strategies, it is essential to understand the underlying mechanisms that lead to failures. Only with a thorough knowledge of failure mechanisms can you develop effective preventive measures and choose the right maintenance strategy.

The understanding of why assets fail is crucial for developing effective preventive maintenance. A bolt can, for example, fail due to loosening, corrosion, fatigue, fracture from overload, or other reasons. To prevent failures, you need to take separate measures for each cause. For bolts that are susceptible to corrosion, it may be useful to replace them regularly or to use a material that is more resistant to rust. For bolts that tend to loosen, it is advisable to improve the tightening procedure or to modify the design.

How long a part lasts therefore varies not only from part to part, but also depending on the cause of failure and the circumstances. The maintenance plan must also be tailored to the specific failure cause and not to the part itself.

6.1. Technical Aging Mechanisms

All technical systems undergo gradual degradation through various mechanisms. This aging follows predictable patterns, but the rate at which it occurs is influenced by operational conditions, loads, and maintenance quality.

Overview of Failure Mechanisms

Loads on technical systems can be divided into several main categories:

Mechanical loads lead to deformation, fatigue, fracture, delamination, creep, and wear.

Thermal loads lead to melting, charring, and thermal expansion and shrinkage.

Chemical loads lead to oxidation, corrosion, and material aging.

Electrical loads lead to aging of insulation material, arcing, partial discharge, dielectric losses, and leakage current.

Cosmic loads (radiation) lead to material aging and polymer degradation.

Degradation Patterns

The condition deterioration of a component can be represented as a gradual decrease in load-bearing capacity. When the condition has declined to the point where the component can no longer withstand its load, failure occurs.

The moment of failure depends strongly on the component type and the specific application. The load is rarely constant — an increase can lead to premature failure.

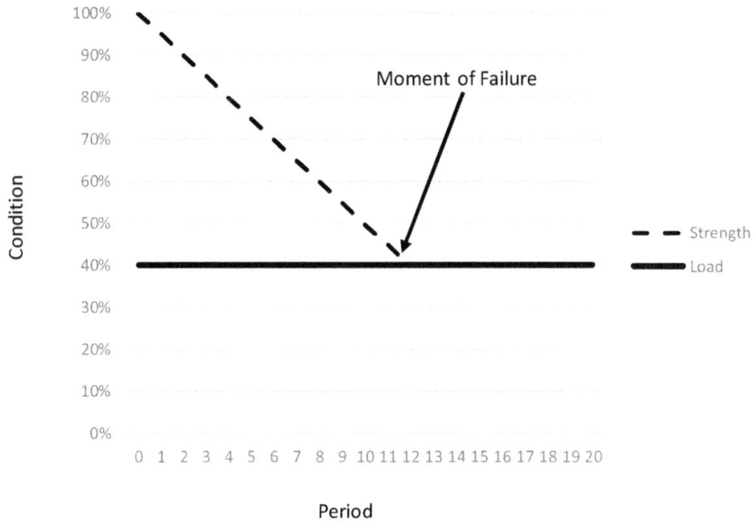

Figure 6-1 Moment of failure under constant load

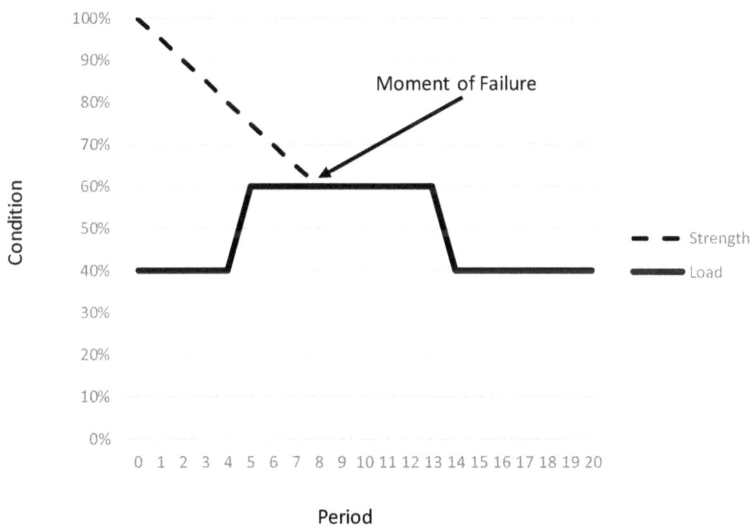

Figure 6-2 Moment of failure under increased load

This heavier load arises from higher ambient temperature, changes in raw material composition, operational factors such as vibrations caused by misalignment, and process variations that lead to peak loads. All these factors together create large differences in degradation between seemingly identical components.

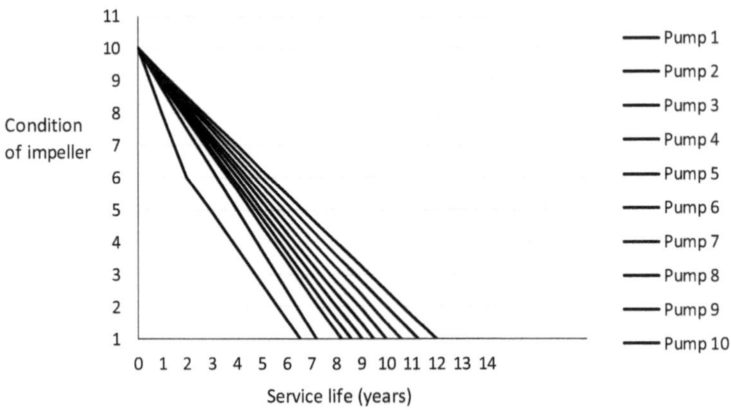

Figure 6-3 Condition degradation

Even when the load that components must be able to withstand is constant, the service life of one component compared to another, identical component can vary enormously. In 90 percent of cases, it is not possible to predict what the service life of a component will be.

Figure 6-3 shows schematically the degradation of ten pumps of the same type. A small variation in degradation rate leads to a large variation in service life for nine of the ten pumps. In the graph I have also included a tenth pump that initially degrades more rapidly for a number of years, for example due to misalignment or cavitation.
After two years, this is corrected and the pump continues to degrade at an average rate. By then, however, the damage is already done and the service life of the pump is shorter than that of the other nine. In the following section, you will read about the background of the spread in service life due to one of the many possible causes of failure: metal fatigue.

Example: Fatigue and Service Life Calculation

For fatigue damage, the Wöhler curve (S-N curve) shows the relationship between stress and service life. For steel, a fatigue limit exists below which the material theoretically lasts indefinitely, while aluminum will always

ultimately fail due to fatigue. Depending on a number of factors, the service life of aluminum can decrease by a tenth (!) power as the stress in the material increases. A doubling of the stress then leads to a reduction in service life by a factor of one thousand.
Reducing the stress in aluminum from 250 MPa to 200 MPa extends the service life, for example, from 60,000 cycles to 500,000 cycles — more than eight times as long.

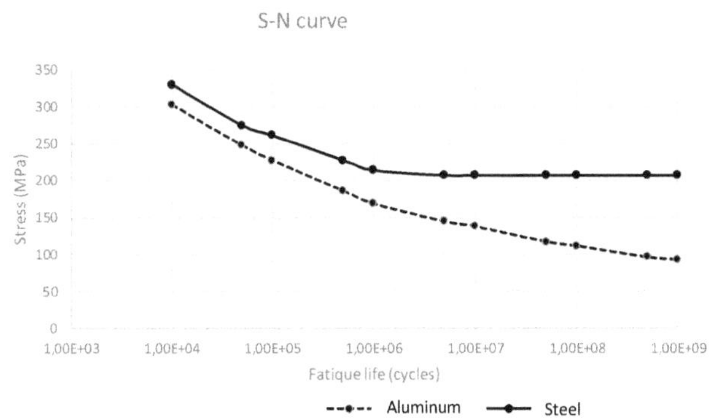

Figure 6-4 Wöhler curve for fatigue

Peak loads during startup can be more decisive than the normal operating load. Suppose that when a machine starts up, the local stress in a steel shaft is 300 MPa. The service life is then 20,000 cycles. Once it is running, the stress may drop to half that, 150 MPa. The service life at that load is theoretically infinite. The shaft therefore only suffers fatigue during startup of the installation. The service life is then determined primarily by the number of starts rather than the number of operating hours.

Example: The Consequence for Bearings

To better understand failure mechanisms, we work out a ball bearing as an example. A ball bearing can fail due to many different causes, such as poor alignment, imbalance, improper lubrication, contamination, moisture, or corrosion. Yet the calculation for the service life of ball bearings is based solely on bearing failure due to surface fatigue.
Even if surface fatigue were the only cause of bearing failure, the spread in service life would still be very large. A small deviation in conditions, the

quality of parts, or the quality of installation can already have a significant impact on the service life of a component.

Many of these kinds of situations occur across all types of technical systems, causing them to fall short of the service life they could theoretically achieve: imbalance, misalignment, exposure to excessive temperature, moisture, contamination, vibration, and excessive load.
These influences cause many components to fall short of their design life. The causes of this are often overloading, improper operation, and inadequate maintenance.

Sarah discovered this at TechPro's motor reducers. "We thought that the bearings simply showed wear," she said. "But by systematically reviewing the history, we saw patterns: machines that were poorly aligned failed after two years, properly aligned and balanced ones lasted more than eight years. It wasn't random wear — it was preventable."

Temperature-Dependent Degradation

Temperature has an exponential influence on most degradation processes. For electrical insulation, doubling the electric field strength can reduce service life by a factor of 500. In metals under high temperature, creep occurs — a temperature increase of 25°C (from 900°C to 925°C) can reduce service life by a factor of three.

These exponential relationships mean that small improvements in operating conditions can yield dramatic increases in service life. A motor that normally operates at 85°C, but can function at 75°C through better cooling, often has twice the service life.

6.2. Spread in Service Life

For many components, it is typical that the period after which ten percent of the components have failed is five times shorter than the period after which fifty percent have failed. The period after which ninety percent have failed is often five times longer again — meaning twenty-five times as long as the period after which the first ten percent have failed.

Replacing components after the lifespan of the first ten percent is therefore unnecessarily expensive, because many good components with considerable remaining service life are then replaced. On the other hand, it is often unacceptable that ten percent of the components have already failed before maintenance is performed.

The service life that a manufacturer specifies for a component is the service life that, for example, 90 percent of components of that type will achieve under specific conditions. These may be laboratory conditions, where the component operates under favorable conditions and is properly maintained. In addition, components rarely operate under perfectly maintained, ideal conditions.
If you plot the lifespan of 1,000 components in a bar chart, you often see the pattern shown in Figure 6-5. Perhaps 80 components already fail in the first month, while some others last as long as 41 months.

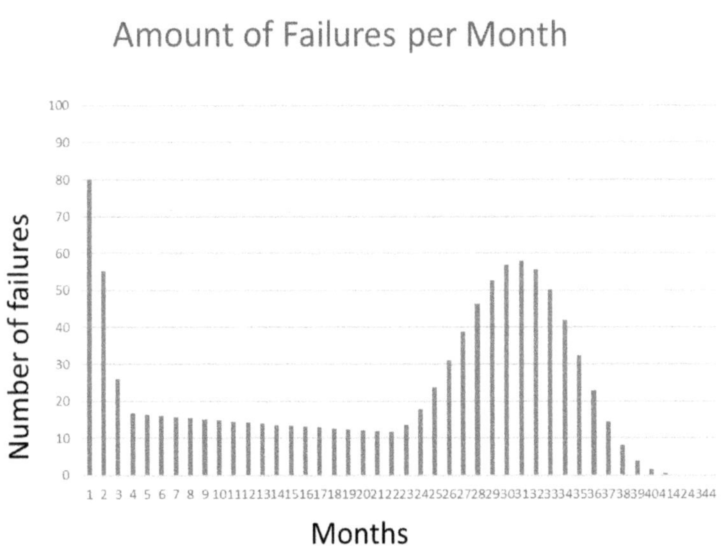

Figure 6-5 Number of failures per month

The failure of these components can also be displayed in a graph like Figure 6-6. This shows the cumulative failures up to and including a specific period. After 41 months, 100% of the components have failed. This graph is called the S-curve.

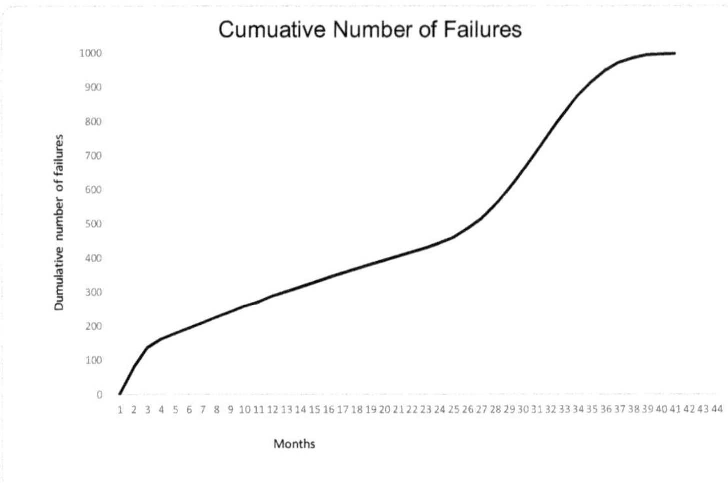

Figure 6-6 The S-Curve

The Swedish mathematician Waloddi Weibull established that if you plot this S-curve on paper with a special logarithmic scale, it usually becomes a straight line.

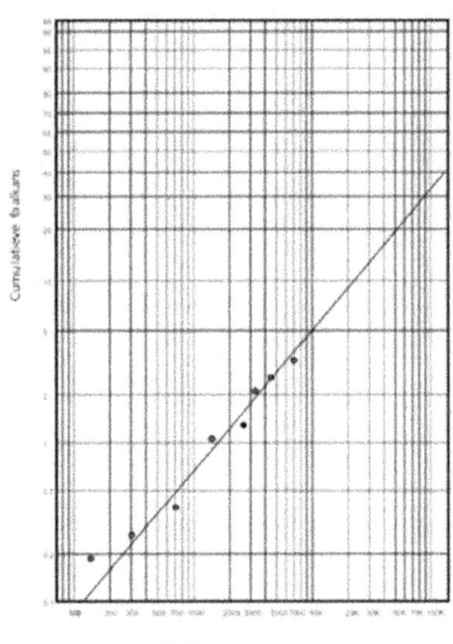

Figure 6-7 The service life of components on a Weibull scale

This allows you to predict, in certain cases, how many components will fail in the future if you have data from the first few periods. The problem, however, is often that this data is unavailable. Companies frequently lack a sufficient number of identical components — which also need to operate under the same conditions and load — to collect meaningful data. If you have two pumps that each fail on average once every ten years, and you need ten data points to fit a curve, you first have to collect fifty years of data before you can apply this data meaningfully. Weibull analysis certainly has a number of useful applications, but it is too complex for the purposes of this book.

Based on the data underlying the previous graphs, you can also create a graph that shows what the probability is in a given month that the component will fail in the coming month. We call this graph a 'failure pattern'.

6.3. The Six Failure Patterns in Practice

Sarah's analysis at TechPro brought a surprising discovery to light. "We thought all components would fail according to the classic wear pattern—performing well for a certain service life, then gradually deteriorating," Sarah said. "But when we systematically analyzed the failure data, we saw six completely different patterns of failure behavior."
Extensive research into failure modes shows that failures develop according to six recognizable patterns, each with specific implications for maintenance strategy.

Pattern A (4% of failure modes): A pattern with increased failures at the beginning, followed by a stable period and rising failures with aging. Do not confuse this pattern with the bathtub curve found in much of the literature. That curve represents the progression of the total number of failures that an asset experiences as a whole over time. It is the sum of all six failure patterns.

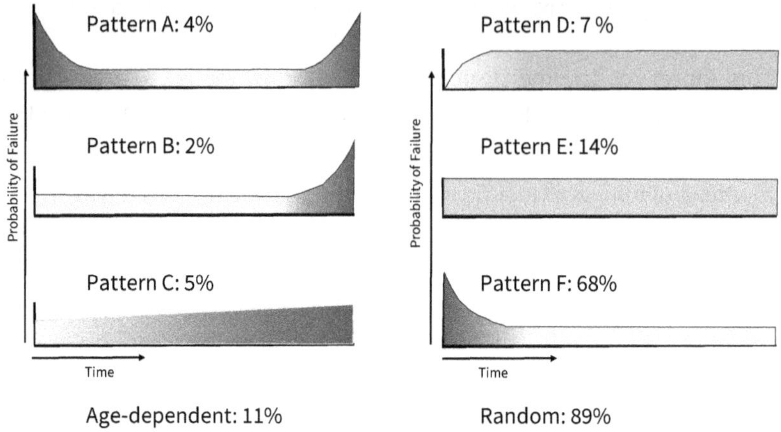

Figure 6-8 The six failure patterns

Pattern B (2% of failure modes): Starts low and flat, then after some time the failure probability begins to rise exponentially. TechPro's corrosion problems followed this pattern: "The steel components in our outdoor installations ran fine for years, but after five years we suddenly saw a sharp increase in rust-related failures."

Pattern C (5% of failure modes): Linearly increasing failure probability throughout the entire service life. With fatigue damage to rotating shafts, Sarah observed this pattern: "Shafts that continuously experienced vibrations showed a steady, predictable increase in failure risk as they accumulated more operating hours."

Pattern D (7% of failure modes): Starts low and rises fairly quickly to a level that subsequently remains constant. "Some of our fast-moving parts in machines follow this pattern — stable at first, then a period of adjustment problems, after which they became stable again at a somewhat higher constant level."

Pattern E (14% of failure modes): Constant, random probability of failure throughout the entire service life. "Human errors, such as incorrect operation or accidentally causing damage, can occur at virtually any time - regardless of the age of the machine. Sudden events such as a stone hitting a windshield or a lightning strike can also happen at any time."

Pattern F (68% of failure modes): Increased probability of failure immediately after installation or maintenance, followed by a rapidly declining probability to a low constant level. This was by far the most common pattern at TechPro. "Most of our problems arose in the first few weeks after assembly or overhaul," Sarah discovered. "Incorrect alignment, installation errors, contamination during maintenance — these kinds of issues accounted for the vast majority of our failures."
The implications for maintenance are far-reaching. With pattern F, replacing after a fixed interval actually increases the probability of failures. "We realized that a lot of our 'preventive' maintenance was counterproductive," Sarah concluded. "We were creating failures instead of preventing them."

For each failure pattern, a different maintenance approach is needed:

- Patterns A, B, and C (combined 11%) can benefit from usage-based maintenance, but often also from condition-based maintenance
- Patterns D, E, and F (together 89%) are random and usage-based maintenance would be useless
- Pattern F requires above all a focus on proper assembly and avoiding unnecessary interventions and condition-based maintenance

6.4. Human Factors

The most common cause of the unpredictable nature of failures is the human factor. Approximately half of all failures are caused by operating or assembly errors, especially in recently installed or overhauled systems.

Types of Human Errors

Assembly errors include incorrect installation of bearings or seals, inadequate alignment of rotating equipment, improper torque of bolts, contamination during assembly, and incorrect cable connections.

Operating errors include exceeding operating limits, incorrect start-up procedures, ignoring warning signals, and the use of wrong materials or settings.

Maintenance errors include insufficient or incorrect lubrication, use of wrong spare parts, inadequate cleaning prior to maintenance, and damage during disassembly.

Control of Human Error

Effective measures include standardized work procedures, adequate training and certification, error-resistant system design, double checks for critical operations, and systematic analysis of incidents.
Sarah's team at TechPro implemented an alignment checklist after every assembly. "Simple but effective," Sarah said. "Ever since we introduced that checklist, early bearing failures have decreased by 70%."

6.5. Strategies for Life Extension

The exponential relationship between load and service life means that small improvements in operating conditions can yield dramatic extensions in service life.

Optimization of Mechanical Systems

For rotating equipment, precision assembly can significantly extend service life.

Alignment and balancing: Even small deviations can dramatically shorten service life. A reduction in vibrations of 20% can double the bearing life.

Better lubrication: Correct lubricants and lubrication intervals prevent premature wear. Over-lubrication can be just as harmful as under-lubrication.

Operating conditions: Controlling temperature, moisture, and contamination has a major impact on service life. Dust shielding and temperature control are often cost-effective measures.

Control of Electrical Components

The lifespan of electrotechnical components is strongly influenced by environmental factors. For many components, a rise of 10°C cuts the lifespan in half.
According to military standard MIL-HDBK-217F, the failure rate varies dramatically. The factor in parentheses in the list below indicates the rate of degradation:

- Controlled environment (baseline factor 1.0)
- Average industrial environment (factor 2.0)
- Mobile applications (factor 9.0)
- Marine environment below deck (factor 7.0)
- Unprotected marine environment (factor 15.0)

Protective Measures

By keeping electrotechnical components clean, dry, and cool, their service life can be significantly extended. There are various measures you can take: keep electrical cabinets closed as much as possible, use cooling or heating to prevent condensation where necessary, create positive pressure to prevent infiltration, install effective dust filters, and use the correct cable glands (preferably on the bottom, so that no moisture can seep in).

Load Factor Management

One of the most effective strategies is reducing the operational load. By operating assets at 70-80% of their capacity instead of 95%, the service life can increase exponentially. This sometimes requires overcapacity in the design, but leads to significant savings in maintenance costs and increased reliability.
A small reduction in the load on a system can extend its service life many times over. Occasional overloading has a very large impact on the service life of equipment, while a small improvement in conditions, such as a temperature reduction, has a very large impact on service life.

6.6. Key Conclusions

All of the previously mentioned factors lead to several conclusions that are very important if you want to establish an effective maintenance program:

- The spread in the service life of components is very large
- In many cases, the spread is so large that it is not possible to predict in advance what the service life of a certain component will be
- Replacing components purely on a time basis leads either to replacing them too early (wasting remaining service life) or too late (failures already occurring)
- Many components do not reach their design life due to overloading, improper operation, and inadequate maintenance.
- Incidental overloading has a very large influence on the service life of equipment
- Human error causes half of all failures
- A small improvement in conditions has a very large influence on service life
- A small reduction in the load on a system can extend its service life many times over

Sarah's research at TechPro confirmed these conclusions. "By improving alignment, applying proper lubrication, and avoiding overloading, we extended the average service life of our gear motors by 60%," she reported. "The investment in precision maintenance paid for itself within a year."
Now that you understand why assets fail and which patterns failures follow, you can apply this knowledge directly when choosing maintenance types. An asset that exhibits pattern F (68% of failure modes) does not benefit from usage-based maintenance, but rather from condition-based maintenance.

After reading this chapter, you may have become somewhat pessimistic about the possibility of establishing a meaningful interval for a maintenance action. Yet this is entirely possible. How you determine this depends strongly on the chosen maintenance strategy. For condition-based maintenance, you determine the interval differently than for usage-based maintenance. For functional tests, the approach is different again. You can read more about determining intervals in chapter 8. In chapter 7, you will first learn how to select the right maintenance approach for each situation.

7. Maintenance Types and Applications

Sarah faced a new challenge. After the successful RCM analysis of pump P-401, she had to implement the right maintenance types for all TechPro assets. "Monthly vibration measurements for critical assets, quarterly inspections for important assets," she noted. But what she found truly exciting was the IoT pilot: wireless sensors on P-401 for continuous monitoring. "If this works, we can predict failures days in advance," she thought enthusiastically.

As you saw in chapter 6, different failure modes follow different failure patterns. This knowledge now helps you choose the right maintenance approach. In this chapter, we cover the four main types of maintenance and their specific applications, so you can select the most effective method for every situation.

7.1. Age-Based Maintenance

According to the RCM decision diagram, usage-based maintenance is only meaningful if the probability of failure increases as more time passes. This occurs in only 11% of failure modes (failure patterns A, B, and C).

When Usage-Based Maintenance Is Effective

Usage-based maintenance works best when a component typically fails due to a single dominant cause that is time-dependent. This happens, for example, with filters that become contaminated at a fairly constant rate and therefore require predictable replacement. For filters that only capture contamination during incidents, periodic replacement is usually not practical.

Wear parts such as brake pads, V-belts, and seals in constant environments often follow predictable wear curves. Catalysts in chemical processes where the catalyst gradually loses its activity are also suitable for this type of maintenance.

Limitations of Usage-Based Maintenance

The most common failure pattern (F, 68% of all failure modes) actually shows an increased probability of failure immediately after new construction or maintenance. For this majority of failure modes, periodic maintenance leads to an increased failure probability due to "infant mortality" of new components.

Even for the 11% of failure modes where usage-based maintenance is applicable, condition-based maintenance is usually preferable, because it makes full use of the entire service life of components, provides insight into actual degradation rates, offers flexibility under changing conditions, and carries a lower risk of installation errors.

Only when practical considerations do not allow for it is usage-based maintenance your best choice.

7.2. Condition-Based Maintenance

Condition-based maintenance is possible when a physical parameter exists that predicts an approaching failure with sufficient certainty. This approach utilizes the full service life of components and reduces unnecessary interventions.

Condition indicators that predict failures

Various indicators can predict impending failures.

For **mechanical indicators,** consider visible wear or degradation, cracks detectable via NDT techniques, vibrations and abnormal sound, temperature changes, and material aging such as hardening or embrittlement.

Performance indicators include energy consumption, capacity or flow rate, efficiency parameters, and quality parameters of the final product.

Chemical indicators are lubricating oil analysis, coolant analysis, and corrosion products.

For **electrical indicators,** you look at insulation resistance, current draw patterns, temperature patterns via thermography, and harmonics in the current waveform.

Implementation Formats

You can observe these indicators in various ways.

Manual inspection means visual or auditory checks performed by operators or maintenance personnel according to checklists.

Periodic measurements are specialized condition monitoring with instruments during scheduled inspections.
Continuous monitoring involves installing permanent sensors for real-time surveillance with automatic alarms.
Your choice depends on the criticality of the asset, the cost of monitoring versus the consequences of failure, and the available technology.

7.3. Inspection Techniques

Sensory Inspection

Visual and auditory inspection are among the most widely used inspection methods. For effective implementation, you must pay attention to several aspects.

Regarding safety aspects, you avoid inspections that require the removal of safety guards while hazards are present. You design inspection points that are safely accessible.

For process management, several rules apply. You assign frequent inspections (weekly or more often) to the production department. You schedule less frequent inspections through your EAM system. You ensure clear procedures for following up on deviations.

For effective checklist development, formulate clear accept/reject criteria for each checkpoint. Use one column on your checklist to indicate whether the checkpoint is in order or not. This allows the work planner evaluating the checklist to see at a glance whether everything is satisfactory. Include part numbers of components to be inspected on your checklist. If someone marks that a component is not in order, the work planner does not need to search through drawings to identify the exact component. Leave room for additional comments – the person performing the inspection may discover things not listed on the checklist. Ensure clear asset identification through QR codes or photographs.

Digitalization of Inspection Processes

Mobile devices linked to EAM systems enable direct recording of findings, access to historical data, trend analysis of degradation, and automatic work order creation upon deviations. This digitization significantly increases efficiency and reduces administrative burdens.

Condition Monitoring with Instruments

The modern trend is shifting from periodic inspection rounds to continuous monitoring through permanently installed sensors. This reduces the required manpower and provides immediate alerts.

Vibration analysis: You measure vibrations on the outside of bearing housings and analyze amplitude and frequency composition. You identify specific problems through characteristic frequency patterns: imbalance shows up at 1x rotational frequency, misalignment at 2x rotational frequency, bearing damage through high frequencies, and gear damage at a frequency that can be calculated from the number of teeth and the rotational speed of a gear. Other problems, such as loose components or poor lubrication, also have their own characteristic vibration patterns.

Ultrasonic inspection: This technique detects high-frequency sounds that are inaudible to humans and requires less specialized training than vibration analysis. You apply it to detect bearing damage or leaking valves by holding the contact probe of the ultrasonic gun against the metal of the bearing housing. Compressed air leaks in piping systems and electrical discharges in switchgear cabinets can be detected by registering the sound that propagates through the air.

Thermographic inspection: Infrared cameras visualize temperature patterns and detect overheated bearings, electrical hot spots, and insulation problems. You get the best results with temperature differences greater than 5°C.

Figure 7-1 Inspection with an infrared camera

Oil analysis: Laboratories analyze samples for oil quality and wear particles. Metal particles provide insight into wear of specific components: iron indicates gears and cylinders, aluminum indicates pistons and bearings, copper indicates bearing material and electrical contacts. Viscosity and acidity indicate oil degradation.

7.4. Non-Destructive Testing Methods (NDT)

For piping systems and pressure vessels, specialized techniques are available that enable you to detect internal defects without damaging the asset.

When to Apply Which NDT Technique

Your choice of a specific NDT technique depends on several factors.

For surface defects in steel, you use magnetic particle inspection because of the high sensitivity and relatively low cost.
For non-magnetic materials such as aluminum or stainless steel, you choose liquid penetrant inspection.

When internal defects are suspected, radiography or ultrasonic testing is necessary. Radiography provides a permanent image, but requires extensive safety measures. Ultrasonic testing is faster and safer, but requires more experience for correct interpretation.
For routine wall thickness measurements in corrosion monitoring, ultrasonic testing is the standard choice due to its speed and reliability.
With heat exchangers containing many tubes, eddy current testing offers the possibility of rapid screening of all tubes.

Cost-Benefit Analysis per Technique

The costs of NDT techniques vary considerably. Visual inspection and wall thickness measurements are relatively inexpensive and quick to perform. Liquid penetrant and magnetic particle inspection require more preparation, but remain affordable for routine application.
Radiography and advanced ultrasonic testing are more expensive due to the cost of the equipment and required expertise. You apply these techniques when the consequences of missed defects are very serious or when other methods provide insufficient information.

7.5. Modern Technologies: IoT and Digital Monitoring

The fourth industrial revolution is transforming maintenance through the integration of digital technologies. Internet of Things sensors, artificial intelligence, and cloud computing create new possibilities for predictive maintenance that support and strengthen traditional approaches.

IoT Sensors for Condition Monitoring

IoT sensors enable continuous monitoring of parameters that previously could only be measured periodically.
Wireless vibration sensors continuously collect data on bearing condition and detect anomalies within hours rather than weeks. Temperature sensors monitor heat development in real time and warn of overheating before damage occurs.
These sensors have become much smaller and less expensive than traditional monitoring systems. A complete vibration sensor with wireless connectivity now costs a few hundred dollars instead of thousands. How

much it costs to install one depends heavily on factors such as the distance to the central recording point and the complexity of the existing control system. Battery life has improved to several years, enabling maintenance-free operation.
Sarah's IoT pilot on pump P-401 was a success. "The wireless sensors detected bearing problems three weeks earlier than our monthly manual measurements would have," she reported. "For an investment of $2,500 we prevented an unplanned repair of $15,000."

Cloud-based data analysis

Modern IoT sensors collect large amounts of data that cannot be processed efficiently on a local level. Cloud platforms provide the computing capacity for advanced analyses and machine learning algorithms that recognize patterns in the sensor data.
These cloud analytics detect subtle changes in the vibration spectrum that are early indicators of bearing wear. By comparing with databases of thousands of similar machines, deviations can be quickly identified and categorized by problem type and urgency.

Predictive Analytics and Machine Learning

Machine learning algorithms analyze historical failure data to identify patterns that precede failure. These algorithms learn which combination of parameters best predicts when a component will fail. Through continuous training on new data, the algorithms continually improve the accuracy of the predictions.

Practical applications include bearing life prediction based on vibration and temperature data, optimization of lubrication intervals based on oil analysis trends, and prediction of motor failure through analysis of current draw patterns.

Implementation Strategy for Digital Technologies

Modern IoT sensors enable continuous monitoring of parameters that previously could only be measured periodically. This technology supports

the RCM methodology by providing better data, but does not replace the systematic analysis of failure modes and their consequences.
Begin with a pilot project on a limited number of critical assets to gain experience. Invest in training your maintenance team and ensure reliable IT infrastructure with adequate cybersecurity. Measure the results against concrete KPIs such as reduction of unplanned downtime and cost savings.

7.6. Function Tests

Function tests are essential for hidden failures — failures that only have consequences when a second failure occurs. This characterizes safety systems and redundant configurations where backup systems only become active when the primary system fails.

Principle of Function Tests

For systems with a primary component and backup, if the primary system fails, the backup automatically takes over. But a defective backup system goes unnoticed until it is needed. Only regular function tests can verify the operational status.
A typical example is an emergency generator that automatically starts when power fails. If the battery of the starting system is dead, the emergency generator cannot start either. This will only become apparent when the generator is actually needed. With a periodic test, you can verify whether all systems are functioning.

Practical Implementation

You develop standardized test protocols that verify all critical functions without unnecessary risks. For emergency generators, for example, you test the automatic start, the transition from grid power to generator, and the switchback to grid power.
When the test fails, you follow clear procedures for immediate corrective actions. A failed backup system must be repaired immediately, because the redundancy is temporarily lost. You inform all stakeholders about the reduced availability during repair.

Sarah implemented a systematic function test program for all safety systems at TechPro. “We discovered that 15 percent of our backup systems would not have functioned in an emergency,” Sarah said. “Now we test everything twice a year and we have 100% certainty.”

Now that you understand why assets fail and what patterns failures follow, you can apply this knowledge directly when choosing maintenance types. An asset that exhibits pattern F (68% of failure modes) has no benefit from usage-based maintenance, but benefits precisely from condition-based maintenance.

In chapter 8 you will learn how to apply the economic optimization of maintenance to every situation in order to determine the most cost-effective approach.

8. Economic Optimization of Maintenance

Now that you have learned the methodologies for determining optimal preventive maintenance and have gained insight into failure patterns and failure mechanisms, we turn our attention to the economic aspects of maintenance decisions. Developing a technically excellent maintenance plan is only the first step — this plan must also be economically sound and fit within your available budget.

Maintenance budgets are under pressure in many organizations, because maintenance ranks among the top three operational costs. At the same time, the costs of failures and unplanned downtime are often even higher than the costs of preventive maintenance. Your challenge lies in finding the optimal balance where total costs (preventive maintenance plus remaining failure costs) are minimized.

8.1. Cost-Benefit Analysis

Identifying Cost Components

An effective cost-benefit analysis begins with fully mapping all relevant cost components. The costs of preventive maintenance are usually clearly visible and include labor costs for internal and external personnel, material and tooling costs, contract costs, and the costs of planned production downtime. These costs are relatively straightforward to determine, because they are directly related to planned activities.

The savings from avoided failures are more complex to quantify. Here you need to look not only at the direct repair costs, but also at all the indirect consequences of failures. Corrective repairs often cost substantially more than preventive actions, because they are unplanned, frequently occur outside of business hours, require diagnostic time, and necessitate rush deliveries of parts.

In addition, failures give rise to costs for unplanned production downtime, potential product rejection if the failure occurs during production, and in some cases customer claims or reputational damage. With critical assets, failures can also lead to consequential damage to other systems, which further increases the total repair costs.

Cost Valuation of Production Downtime

One of the most complex aspects of maintenance economics is the valuation of production downtime. The cost of one hour of downtime is not constant but depends on various factors such as available overcapacity, the timing of the failure, and the possibilities for make-up production.
In organizations with excess capacity, short outages can often be compensated by running harder later or through weekend production. In such cases, the actual cost of downtime is lower than the theoretical production loss. Conversely, in organizations running at full capacity, outages can lead to delivery delays, customer claims, and loss of revenue.

For practical applications, it is useful to distinguish between different categories of downtime costs. Short stoppages (up to 30 minutes) often have limited impact, because they can be made up within the same shift. Stoppages of up to two hours can possibly be compensated within the same week. Longer stoppages often require extra shifts on the weekend or result in actual production loss.
The concrete methodology for performing a complete cost-benefit analysis is illustrated in section 8.2 with the example of replacing V-belts, where all cost components are calculated step by step.

8.2. Determining Optimal Intervals

One of the questions managers ask most frequently about preventive maintenance is: "How often should we perform the maintenance?" The answer to this question depends on many factors. In this section, you will learn how to determine the optimal interval for the various preventive maintenance strategies.

Usage-Based Maintenance

Failure modes that follow failure pattern A, B, or C (11% of all failure modes) can be effective candidates for usage-based maintenance. You determine the optimal replacement interval by balancing preventive and corrective costs.

Methodology for Interval Calculation

The total costs per time unit consist of two components: the preventive costs and the expected corrective costs. By combining both cost components, a total cost curve is created whose minimum indicates the economically optimal interval.

The preventive costs per year are calculated as:

Annual preventive costs =

$$\text{Cost per maintenance action} \div \text{Interval in years}$$

The preventive costs per year are halved if you double the interval between two maintenance actions. You then get a graph like figure 8-1.

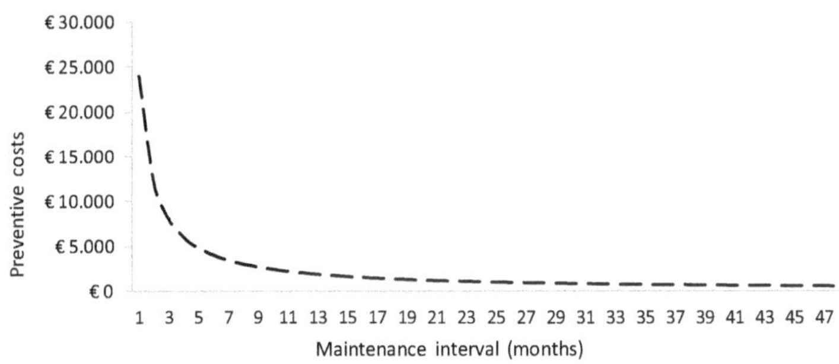

Figure 8-1 Annual costs of preventive maintenance in relation to maintenance interval

The expected corrective costs are proportional to the probability of failure, which typically follows a Weibull distribution. This is not linear — when the interval is doubled, the probability of failure increases more than proportionally.

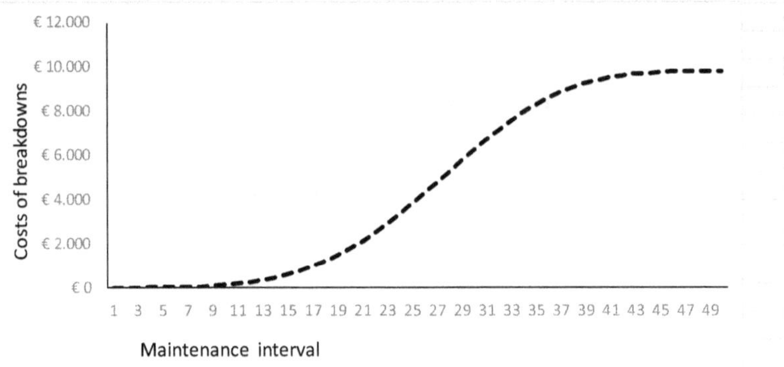

Figure 8-2 Annual failure costs in relation to maintenance interval

It is important when calculating failure costs to include not only the costs of labor hours and parts for the repair, but all costs. The largest cost component is often failure-dependent costs, such as the costs of production loss.

The optimal maintenance interval is the interval at which the sum of preventive costs plus failure costs is minimal, as Figure 8-3 shows.

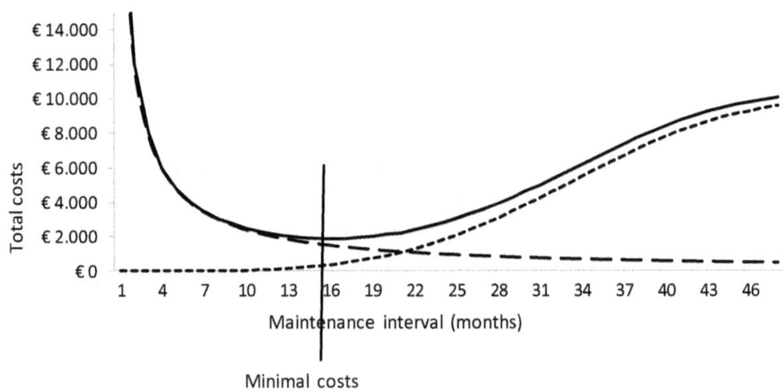

Figure 8-3 Total maintenance costs in relation to maintenance interval

Practical Example:

Sarah's plant has 120 fans with V-belts that are essential for the cooling process. These belts could in principle be inspected for condition-based maintenance, but, because her company rarely has planned shutdowns and most fans run continuously, a usage-based strategy is more practical.

Given data:

Preventive replacement costs: $45 per set of belts (material $25, labor $20)

Corrective repair costs: $180 per set (material $25, diagnosis $15, rush $20, extra labor $45, production downtime $75)

Historical failure data from comparable belts shows: 8% fail after 1 year, 25% after 2 years, 50% after 3 years, 75% after 4 years, 90% after 5 years

Calculation of optimal interval:

Interval 1 year:
Annual preventive costs: $45 ÷ 1 = $45 per set
Expected annual failure costs: 0.08 × $180 = $14 per set
Total annual cost: $59 per set

Interval 2 years:
Annual preventive costs: $45 ÷ 2 = $23 per set
Expected annual failure costs: 0.25 × $180 = $45 per set
Total annual cost: $68 per set

Interval 3 years:
Annual preventive costs: $45 ÷ 3 = $15 per set
Expected annual failure costs: 0.50 × $180 = $90 per set
Total annual cost: $105 per set

Interval 4 years:
Annual preventive costs: $45 ÷ 4 = $11 per set
Expected annual failure costs: 0.75 × $180 = $135 per set
Total annual cost: $146 per set

Conclusion: The optimal replacement interval is 1 year, where total annual costs are lowest ($59 per set). With 120 sets of belts, this means an annual savings of $1,080 compared to a 2-year interval.

Additional Benefits

- Fewer unplanned downtime events during critical production periods
- Better ventilation efficiency (new belts have less slip)
- Lower risk of secondary damage to motors from overloading

Condition-Based Maintenance

The P-F Curve and the P-F Interval

With condition-based maintenance, you need to align the inspection frequency with the P-F interval: the time span between the detection of a potential failure (P) and actual failure (F).

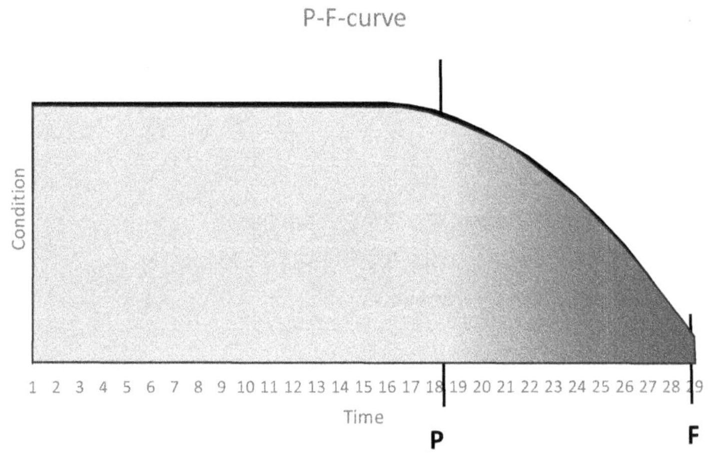

Figure 8-4 The P-F curve

Concept of the P-F curve: The P-F curve shows how the condition of a system deteriorates over time. Initially, quality remains stable within normal fluctuations. At a certain point, the deterioration becomes measurable — this is point P (Potential failure). From this point, the condition continues to worsen until the system actually fails at point F (Failure). The time between point P and point F is called the P-F interval.

It is important to understand that the P-F interval is not determined per component, but per failure mode. A bearing, for example, can have multiple P-F intervals: wear from normal loading may have a P-F interval of three months, while insufficient lubrication has only one month or sometimes even just a few minutes P-F interval. Alignment problems can have yet another interval of six weeks.
This means that you must align your inspection frequency with the shortest relevant P-F intervals if you want to detect all significant failure modes. Monthly vibration monitoring detects both slow wear and many rapid lubrication problems. Quarterly inspection may miss the faster degradation processes.

For practical applications, this means that you cannot be satisfied with identifying "the" P-F interval of a component. You must analyze the different failure modes and determine what percentage of failures you want to detect within what timeframe.

Determining the P-F interval: The inspection frequency must be shorter than this interval. If you inspect just before point P, everything still appears to be in order. If the next inspection doesn't take place until after point F, you are too late.

Factors that influence the P-F interval:

- The component: Bearings have different P-F intervals than belts or pumps
- Circumstances: Temperature, load, and environmental conditions
- The failure mode: Wear, fatigue, or corrosion have different rates of development
- Inspection technique: Vibrations can detect wear months earlier than visual inspection

Practical Inspection Guidelines:

Vibration Measurements:

- Standard equipment: quarterly
- Critical systems: monthly

- Highly critical systems (turbines): continuous monitoring

Thermographic inspection:

- Rotating equipment: weekly to biweekly
- Electrical systems: semi-annually (critical: quarterly)
- Insulation/heating: annually before winter season

Oil analysis:

- Standard: three to six months
- Depending on load and contamination level

From the P-F interval, you subtract the time required for planning and preparation. If a P-F interval is 3 months, but you need 1 month to prepare a planned shutdown, then the optimal inspection interval is 2 months.

Do not unnecessarily shorten the inspection interval — this adds no value. Do not extend the interval either if inspections always indicate 'good,' because some failures can only be detected a very short time in advance.

Cost Optimization in Condition-Based Maintenance

With condition-based maintenance, the P-F interval largely determines the inspection frequency. The economic optimization lies not so much in the interval itself (which is determined technically), but in the question of whether it is cost-effective to perform this condition monitoring at all.

Cost components:

- Inspection costs: Dependent on number of assets and inspection frequency
- Preventing failure costs: Savings through early problem detection
- Remaining failure costs: Some failures are still being missed

The total cost of condition monitoring must be lower than the expected failure costs without monitoring. If a vibration measurement costs $50 per motor per quarter, but prevents $300 in failures annually, the program is cost-effective.

The advantage of condition-based maintenance is that you only intervene when necessary, thereby avoiding unnecessary preventive actions and allowing the actual condition to drive maintenance decisions.

Function Tests

How often you need to perform a functional test depends on a number of factors. You determine the test frequency by first establishing, in consultation with management, how often the entire system is allowed to fail. You can then calculate the test frequency using the formula:

$$T(FT) = (Tb \times Th \times 2) / Ta$$

With:

T(FT)	=	the interval between functional tests.
Tb	=	the Mean Time Between Failure of the backup system
Th	=	the MTBF of the main system
Ta	=	the acceptable interval between failures of the entire system

For a system that may fail at most once every 200 years, with a backup MTBF of three years and a main system MTBF of 20 years, the following applies:

$$T(FT) = (20 \times 3 \times 2) / 200 = 0.6 \text{ years.}$$

Testing the system once every 7.2 months is therefore sufficient.

8.3. Investment Decisions

Sometimes a one-time modification is more cost-effective than repeated preventive maintenance. In the context of preventive maintenance, this situation arises when an asset has structural problems that can only be resolved by adapting the design, improving the installation, or replacing it with a more reliable version.

When Do We Consider Modifications?

Modifications come into consideration when recurring maintenance problems cannot be effectively resolved with preventive actions. Examples are:

- **Design flaws:** A pump that is structurally overloaded
- Environmental problems: Bearings that fail repeatedly due to contamination
- **Obsolete technology:** Components for which parts are no longer available
- **Inefficient systems:** Assets that require far more maintenance than modern alternatives

Advantages and Disadvantages of Modifications

Modifications offer the primary advantage of being a one-time solution — after a successful modification, the recurring maintenance problem disappears permanently. In addition, you no longer run the risk at every maintenance interval of something being forgotten or going wrong, which eliminates execution risk. Modern solutions are also generally more efficient and reliable than the original design, which often leads to improved performance.

Against these advantages, however, there are also disadvantages. Modifications often require substantial one-time costs, meaning a higher initial investment than repeated preventive maintenance. In some sectors, a modification requires much more time due to verification and approval procedures, leading to longer lead times. The greatest risk, however, lies in the design aspect: poorly thought-out modifications can introduce new problems — you solve one problem and get two back in return.

Simple Evaluation Method

For maintenance-oriented investment decisions, a simple calculation of the payback period is often sufficient:

Payback Period = Investment Costs ÷ Annual Savings

As a rule of thumb: modifications are worthwhile if the payback period is shorter than 3–5 years.

Practical Example

A conveyor belt has chronic problems with drive belts that need to be replaced every six months due to poor alignment of the drive rollers.

- Annual maintenance costs: 2 × $450 = $900 per year
- Modification (drivetrain redesign): $3,600 one-time
- Payback period: $3,600 ÷ $900 = 4 years
- Expected service life modification: 12 years

Conclusion: The modification is cost-effective, because the payback period (4 years) is significantly shorter than the expected lifespan of the solution. For more complex investment evaluations with multiple cash flows over longer periods, more complex calculations need to be performed, but for most maintenance-oriented modifications this simple approach is sufficient.

Decision Process

1. Identify the structural problem — Is preventive maintenance really the right solution?
2. Identify modification options — What alternatives are available?
3. Calculate annual savings — How much maintenance is eliminated?
4. Estimate modification costs — Including engineering and implementation
5. Evaluate risks — What could go wrong with the modification?
6. Decide based on payback period and risk — Does it fit your maintenance strategy?

This systematic approach helps you determine when a modification is a better strategy than repeated preventive maintenance.

8.4. Practical Approach for Asset Categories

As described in chapter 4, you categorize assets based on criticality. Instead of analyzing each asset individually, you apply template-based approaches per asset category.

Practical Case: 720 Gear Motor Units

Sarah's experience at TechPro illustrates this approach well. For a production plant with 720 motor reducers, the following categorization was applied:

20 critical motor reducers (score 4-5): • Monthly vibration measurements • Oil change every three years + laboratory analysis • Weekly visual inspection by operators

30 important gear motors (score 2-3): • Quarterly vibration measurements • Oil change every three years (no analysis) • No weekly inspection

670 standard gear motors (score 1):
• No preventive maintenance
• Failure-based maintenance
• Replacement upon failure
This approach saved Sarah's team enormous amounts of analysis time, while resources were optimally deployed on the assets that make the greatest difference.

8.5. Sustainability Aspects of Economic Optimization

Alongside traditional economic considerations, sustainability objectives are playing an increasingly important role in maintenance decisions.

Life Extension vs. Replacement Strategy
Preventive maintenance can significantly extend the service life of assets, contributing to circular economy objectives. An asset that lasts 20 years instead of 15 years due to proper maintenance reduces the need for new production by 25%.

Well-maintained assets often consume 10-15% less energy than poorly maintained assets. A motor running at optimal condition has a higher efficiency than a motor with increased resistance due to faulty bearings or contaminated cooling.

Energy Savings Through Well-Maintained Assets

Sarah discovered at TechPro that well-maintained compressors consumed 12% less energy. "That saves not only $18,000 per year in energy costs," she reported, "but also contributes to our CO2-reduction target." These sustainability benefits can often be included as added value in your business case, which strengthens the economic argument for preventive maintenance further strengthened.

Now that you have learned how to perform economic analyses and determine optimal

when determining maintenance strategies, you can look at chapter 9 for the practical implementation and execution of your maintenance plan. You will learn how to turn individual maintenance actions into a workable overall schedule and how to ensure that the plan is actually carried out, with all the challenges and setbacks that Sarah encountered at TechPro.

9. Implementation and Execution

Sarah was ready for the next step in improving preventive maintenance. All analyses were complete, the business case approved, but now she had to actually implement her preventive maintenance program at TechPro. "Theory is one thing," she sighed, "but practice is stubborn."
The first month didn't go according to plan: the EAM system had limitations, the team resisted change, and some external suppliers couldn't deliver right away. But Sarah held firm, and after six months the transformation was complete: from 70% reactive to 70% preventive maintenance.

Now that you have established which preventive maintenance is technically necessary and economically justified, the crucial phase of implementation and execution begins. Even the best-designed maintenance plan fails if it is not implemented correctly. The transition from theory to practice requires careful planning, clear work instructions, and effective organization of all parties involved.
In this chapter, we cover the practical aspects of developing a workable master schedule, creating effective work instructions, and ensuring proper execution. We give special attention to the challenges that arise when combining many individual maintenance actions into a coherent and efficiently executable program.

9.1. Creating a Master Schedule

From Individual Actions to Workable Units

The result of an RCM or iRCM analysis is a comprehensive list of all maintenance actions that need to be performed on a system. Table 9-1 shows an example of a small portion of such a list.

Asset no.	Asset	Maintenance action	Interval (weeks	Shutdow n?	By
125.1	Air dryer	Check chain for rust deposits	4	Yes	Maintenance
101	Sanding machine	pressure bar (correct control per lifting magnet, spring mechanism operation, ...)	13	Yes	Maintenance
395	Take-off device	Periodic lubrication of bearings	13	No	Production
393	Conveyor belt	conveyor belt; report if replacement is needed	4	Yes	Production
673	Conveyor belt	conveyor belt; report if replacement is needed	4	Yes	Production
104	Epibrating machine	Visual inspection (soft foot, leaks, ...)	4	Yes	Maintenance
110	UV dryer	Clean and inspect motor and fan	4	Yes	Production
110	UV dryer	Ultrasonic inspection	13	No	Maintenance
131	Epibrating machine	Vibration analysis	4	No	Trilco
127	Epibrating machine	Vibration analysis	4	No	Trilco
114	U-former	Check the coupling for fractures, cracks, dents, ...	4	Yes	Maintenance
127	U-former	Check chain for rust deposits	4	Yes	Maintenance
106	Brush machine	Visual check for cracks, fractures, ...	1	Yes	Production
117	Epibrating machine	Ultrasonic inspection	4	No	Maintenance
101	Sanding machine	Infrared inspection	4	No	Maintenance
050	Pad plate station	Ultrasonic inspection	4	No	Maintenance
395	Take-off device	Visual check of cylinder operation	1	No	Production
104	U-former	Ultrasonic inspection	4	No	Maintenance
108	UV dryer	Check valve for contamination / Clean the valve	26	Yes	Production
181	Epibrating machine	Clean and inspect motor and fan	4	Yes	Maintenance
391	Stacking device	Periodic oil replacement	156	Yes	Maintenance
133	Rocker	Compare with the set value. Adjust if necessary.	13	No	Production
050	Pad plate station	Periodic level check	1	No	Maintenance
109	Epibrating machine	Check chain for rust deposits	4	Yes	Maintenance
125.2	Air dryer	Vibration analysis	4	No	Trilco
131	U-former	Ultrasonic inspection	4	No	Maintenance
157	U-former	Vibration analysis	4	No	Trilco

Table 9-1 Long list of maintenance tasks

For a complete production department, this list can contain thousands of lines, with each inspection point, lubrication point, or measurement location forming a separate line. This detailed list is essential for completeness, but presents a practical problem for execution.

It is not efficient to create a separate work order for each individual inspection point. This would lead to a heavy administrative burden and would make efficient clustering of activities more difficult. A technician who checks one temperature sensor can perhaps also check all other sensors in the same area during the same visit.

The art lies in intelligently grouping maintenance actions into workable units that are large enough to minimize administrative overhead, yet small enough to be completed within a reasonable time by one person or team. A work order should preferably be completed in full during a single visit, whether during normal operations or during a planned maintenance shutdown.

Systematic Grouping Approach

An effective grouping strategy follows a logical sequence in which various criteria are applied successively. The first and most important distinction is between actions that require downtime and actions that can take place during operation. This fundamental separation largely determines when and how the maintenance can be performed.

Actions that require downtime are then grouped by interval. It is economically attractive to combine, for example, all monthly downtime actions, so that the production line only needs to shut down once a month instead of multiple times. This often requires adjustment of intervals: an action that actually needs to occur every six weeks might be adjusted to monthly to enable synchronization.

Within each interval group, you further sort by executing party. Work that must be performed by external specialists you combine as much as possible to save on travel costs. A specialist for vibration measurements can, for example, measure all critical machines in a plant in a single day instead of making a number of separate visits.

For work to be performed internally, group by trade discipline and location. Combine all electrical work on a single production line, as well as all mechanical actions. This enables efficient deployment of specialized personnel and reduces the number of different tool sets that need to be brought along. Also ensure that mechanical, electrical, and other maintenance actions are performed simultaneously as much as possible.

Practical Example: Grouping Strategy for a Production Line

Sarah's experiences at TechPro perfectly illustrate this approach. For the packaging line with conveyor belts, packaging machines, labeling machines, and palletizing equipment, she developed a systematic grouping strategy.

Step 1: Shutdown versus operation — First she distinguished between tasks that required a shutdown and those that could take place during operation. Lubricating accessible bearing points, visual inspections, and non-invasive measurements could be done during operation. Work on moving parts required a shutdown.

Step 2: Interval Grouping - For shutdown work, she identified several groups: weekly quick checks, monthly thorough inspections, quarterly precision alignments, annual complete overhauls. By combining all monthly actions, a work order was created that required 4 hours of downtime, but covered all necessary activities.

Step 3: Capacity Planning - For work during operations, she divided the workload: 100 lubrication points were distributed over 4 weeks with 25 points per week, which prevents peak loads.

9.2. Level Differentiation in Instructions

After grouping maintenance actions into work orders, you need to prepare detailed work instructions for the technicians. The level of detail required depends heavily on the knowledge and experience of those who will perform the work. An experienced technician requires different information than

For skilled technicians, a concise checklist is often sufficient, listing what needs to be checked and what specific points of attention apply. For less experienced workers, more detailed instructions are necessary, including step-by-step descriptions, safety procedures, and photos or videos for clarity.

Work Instruction

Location	Line	Machine		OPL: no.	3.25
Powderplant	3	Hammer mill 2		Rev.	2
				Date	6-9-2025

Subject	**Measuring V-belt tension**				
Objective	prevent excessive belt wear				
				Prepared by	Approved by
Safety		Improvement		By	by
Basic knowledge	V	Issues		JS	MvdH

Safety warning: Perform LOTOTO on the machine, before measuring belt tension

Tools: Pencil gauge

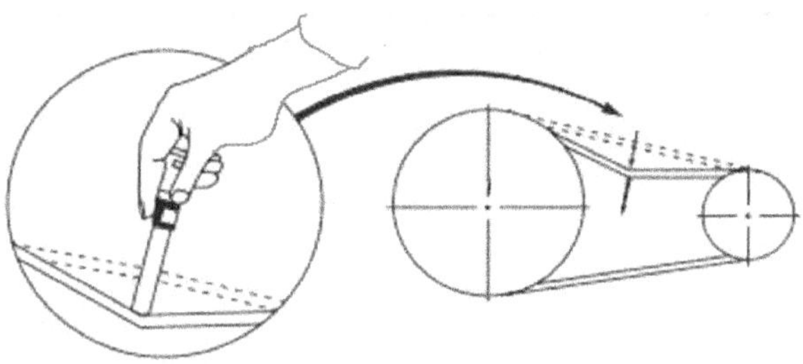

Execution

1. SET THE PENCIL GAUGE
Set the large O-ring to the required deflection: 18 mm
Set the small O-ring on the plunger to zero (0)
2. PERFORM THE MEASUREMENT
Place the gauge exactly at the midpoint of the belt
Hold the gauge perpendicular to the belt
Press the plunger until the large O-ring is level with the adjacent belt
Read the force on the scale (small O-ring): should be 7–9 Newton

Important: Measure both belts – tension must be equal!
Report the result to your supervisor if it deviates from the standard.

Figure 9-1 Work instruction in the form of a One Point Lesson

Practical Example: Work Instruction Vibration Measurement

Sarah developed different versions of work instructions depending on the experience level.

Version for experienced specialist: Perform vibration measurement on pump P-401 according to procedure VIB-001. Measurement points: bearing DE horizontal/vertical/axial, bearing NDE horizontal/vertical. Acceptance criteria per ISO 10816 Class II. Record deviations >4.5 mm/s RMS.

Version for operator with basic training: Detailed step-by-step instruction with preparation, measurement procedure, acceptance criteria, and documentation requirements, including photos of measurement points and clear action thresholds.

Standardization and Consistency

For TechPro's many similar assets, Sarah developed standardized work instructions. By defining standard task lines such as "inspect V-belt," consistency between work orders was achieved and instructions were drafted more efficiently.
Modern EAM systems support building instructions from standard building blocks. A work order for "Quarterly maintenance centrifugal pump" is automatically assembled from standard tasks such as "Check bearings," "Inspect coupling," and "Verify alignment." An instruction for an operator can look like Figure 9-2. Such an instruction is also called a One Point Lesson (OPL).

9.3. Action Items and Safeguarding

Implementation of Supporting Measures

An RCM analysis not only produces a preventive maintenance plan, but also identifies various supporting measures that are necessary for successful implementation. You must address these action items systematically in order to realize the full potential of the maintenance plan.

Technical improvements can range from minor adjustments to substantial modifications.
Training needs arise because new maintenance techniques are being introduced.

Adjustments to operating procedures may be necessary to increase effectiveness.

Spare Parts and Tools

Sarah discovered that implementation had direct consequences for required spare parts and tools. Components that had previously been replaced on failure now needed to be available preventively. New maintenance techniques such as vibration measurements required specialized equipment.
She weighed the investment in measuring equipment against outsourcing to specialists. For TechPro's many identical assets, purchasing their own equipment was cost-effective.

Contract Management and External Parties

Many maintenance activities were carried out by external parties.
Sarah's experience demonstrated the importance of clear performance requirements and acceptance criteria. Vague descriptions led to misunderstandings; specific requirements such as "measure vibrations at 12 defined points according to ISO 10816" provided clarity.

9.4. Common Problems and Solutions

Sarah's implementation experience at TechPro brought several practical challenges to light that were difficult to foresee in advance.

Problem 1: Declining Management Commitment

Situation: After three months, management began to doubt whether the initial investment was justified, as the results were not yet fully visible.

Solution: Sarah implemented a quick wins strategy by focusing on a few visible improvements and regular progress reporting with concrete figures on avoided failures.

Problem 2: Asset Register Turns Out to Be Incomplete

Situation: During implementation, it became clear that the asset register contained more gaps than expected, with new machines not registered and inconsistent numbering.

Solution: Iterative approach in which the register was gradually completed, and application of the 80/20 rule—focusing on the 20% of assets that cause 80% of the problems.

Problem 3: Teams Don't Have Time for Sessions

Situation: Production personnel had difficulty freeing up time for RCM sessions due to operational pressure.

Solution: Shorter sessions of two hours instead of three hours, better preparation by the facilitator, and scheduling during less busy periods.

Problem 4: EAM System Limitations

Situation: The existing EAM system could not support all the desired functionalities for the new maintenance program.

Solution: Looking for workarounds within the existing system and using external tools for specific functionalities such as vibration data analysis.

9.5. Implementation Case: 6-Month Roadmap

Sarah's systematic approach at TechPro follows a proven 6-month roadmap:

Month 1: Preparation and Team Formation — Sarah assembled the implementation team and arranged training in RCM methodology. The team began with criticality ranking of all 200 assets.

Month 2: Criticality Assessment - In 3 day-long sessions, the team performed criticality ranking for all assets. Result: 25 critical, 75 important, 100 standard.

Month 3: RCM Analyses of Critical Assets - Full RCM analyses were conducted for the 25 critical assets in 15 3-hour sessions. In parallel, standard maintenance schedules were developed for other assets.

Month 4: Schedule Development - All individual actions were combined into workable work orders and scheduled in the EAM system. Work instructions were drafted and external contracts established.

Month 5: Training and Pilot - Personnel received training in the new work methods. A pilot on one production line tested the procedures and gathered feedback for improvements.

Month 6: Full Implementation - The new program was rolled out to all production lines with daily KPI monitoring.

Results After Implementation

Sarah's results after six months speak for themselves: 40% less unplanned downtime, 25% lower maintenance costs, and a fully transformed organization from 70% reactive to 70% preventive maintenance.

"The most important thing was perseverance during the difficult start," Sarah reflected. "The first month a lot went wrong, but by systematically solving problems and bringing the team on board, we ultimately achieved fantastic results."

Organizational Assurance

For sustainable implementation, Sarah ensured adequate organizational embedding: documenting responsibilities, training key personnel, and integrating the new maintenance plan into existing business processes.

The role of the reliability engineer remained important after implementation for monitoring performance, analyzing deviations, and proposing improvements. Training of planners and supervisors was crucial, as they had to work with the new plan on a daily basis.

Now that you've learned how to successfully implement and execute your maintenance plan, including all the practical challenges that Sarah encountered, the question arises: can modern technologies further strengthen this systematic approach?

Sarah discovered that IoT sensors, AI analytics, and cloud platforms made her proven RCM methodology more effective without replacing the fundamentals. In the next chapter, you will learn how digital technologies strengthen preventive maintenance — not as a magic solution, but as a logical extension of the systematic approach you now have mastered.

10. Digitizing Maintenance Systems

Sarah looked at the alarm on her phone. 3:17 in the morning. Pump P-401 had exceeded a critical vibration threshold. A year earlier, this would have been the start of a crisis: diagnosing the problem, ordering parts, 8 hours of production downtime, $15,500 in damages. Now she calmly walked to her laptop and opened the dashboard.

The wireless sensor showed a gradual increase in bearing vibrations over three weeks — far too slow to be noticed during their monthly manual measurements, but clearly visible in the continuous data. Sarah scheduled a work order for the following morning:

replace bearing lubrication, 2 hours of work, no production downtime. “This is why we’re digitizing,” she thought with satisfaction.

When Sarah had begun TechPro's maintenance revolution six months earlier, she had no idea that digital technologies would ultimately strengthen her systematic RCM approach. Now, with a fully implemented preventive maintenance program and proven results, she discovered how IoT sensors, AI analytics, and cloud platforms made her methodology more effective without replacing the foundations.

10.1. From Solid Foundation to Smart Reinforcement

Why Foundations Come First

Sarah's digital journey illustrates why organizations must first lay their maintenance foundation before investing in advanced technologies. "Before those sensors, I first needed six months to know which assets were critical and why they were failing," she reflected. "Digital tools amplify systematic thinking, they don't replace it."

The methodology in this book remains the foundation. Digital technologies make each step more effective, but cannot skip a step.

Organizations that reverse this order — technology first, then systematization experience disappointment. Sensors on assets whose failure modes you don't understand generate data without insight. AI algorithms without quality data produce unreliable predictions. Cloud platforms without clearly defined processes create digital chaos.

The Reinforcement Principle

Digital technologies enhance preventive maintenance in four ways: better problem detection through continuous monitoring, more accurate predictions through pattern recognition, more efficient planning through automated workflows, and deeper insights through data analysis over longer periods.

This reinforcement builds on RCM logic. You measure parameters that RCM analyses have identified as predictive of failure. You use AI to recognize patterns in failure modes that you already understand. You automate processes that you have proven manually.

10.2. IoT sensors

From Periodic to Continuous

Sarah's RCM analysis of pump P-401 had identified vibration as the best early indicator of problems. Her original plan: monthly manual measurements. Effective, but limited by measurement frequency and human availability.

IoT sensors eliminate these limitations through 24/7 monitoring. For P-401, Sarah invested $1,600 in a wireless sensor with cloud connectivity. The sensor detects increases in vibration that are too gradual for monthly measurements, but crucial for predicting bearing problems.

Strategic Sensor Selection

After the success with P-401, Sarah expanded to other critical assets. Her selection criteria reflect a systematic approach: start with assets where RCM analyses have identified failure-predictive parameters, choose proven sensor technologies that function reliably in industrial environments, and focus on parameters where manual measurement is logistically challenging.

For TechPro, this meant temperature and vibration sensors on critical rotating equipment, but no sensors on assets where visual inspection suffices. "We don't digitize everything," Sarah explained. "Only where it provides real added value."

Realistic ROI Expectations

Sarah's business case for IoT shows realistic expectations. She compares the investment of $1,600 per critical asset with the cost of a single unplanned failure: $15,500. "If the sensor prevents one failure every two years, it pays for itself. But the real value lies in better planning and less stress for the team."
This calculation illustrates the amplification principle: IoT makes existing maintenance strategies more efficient; it does not replace them entirely.

10.3. Artificial Intelligence: Recognizing Patterns

From Data to Predictions

After a year, Sarah's sensors had collected millions of data points from fifteen critical assets. Too much for manual analysis, but ideal for machine learning algorithms that can recognize subtle patterns.
Sarah is experimenting with AI software that predicts when motors will develop bearing problems. The algorithm combines current consumption, vibration, and temperature data to identify patterns that provide warnings two weeks earlier than traditional methods.

The system discovered three interesting correlations that people had missed: motor M-302 shows specific current draw patterns that precede bearing failure, pump P-205 has temperature fluctuations after weekend shutdowns that predict lubrication problems, and compressor C-101 exhibits vibrations that correlate with outdoor temperature due to thermal expansion.

AI Implementation with Realistic Expectations

Sarah's experience shows both the possibilities and limitations of AI in maintenance. After eight months of training, her system achieves 70% accuracy in predictions — better than their old schedules, but not a perfect crystal ball.

"AI tells us when a bearing is likely to fail, but not whether replacement is economically justified right now," Sarah acknowledged. "That trade-off remains an engineering judgment based on RCM principles."
The most important insight: AI amplifies human expertise by identifying patterns that remain invisible to humans. The decision-making about what to do with them remains a combination of data insight and engineering experience.

10.4. Cloud Platforms: Computing Power and Integration

From Local Limitations to Scalable Solutions

Sarah's growing collection of sensors generated more data than TechPro's local systems could process. Cloud platforms provide the storage capacity and computing power for advanced analyses without major local investments.
For $50 per asset per year, TechPro gets access to cloud storage, basic analytics, and automatic software updates. “Much cheaper than our own data center, and we get features that we could never develop ourselves,” Sarah said.

Integration Challenges and Solutions

The biggest implementation challenge was the integration between cloud-based sensor data and TechPro's local asset management system. Sarah's team had to develop API connections to automatically create work orders based on sensor alarms.
"Our work orders come from the local system, but sensor data from the cloud. Getting those two to talk to each other took six months instead of three," Sarah said. This experience illustrates why digital transformation requires more IT expertise than many organizations realize.
For smaller organizations, fully cloud-based solutions can avoid this integration complexity, but do require trust in external data security.

10.5. Implementation Strategy

Sarah's Phased Digital Journey

Sarah's approach shows how organizations can realistically experiment with digital technologies without risking their operational stability. After completing her preventive maintenance program, she started a new phased implementation for digitalization.

Phase 1: One Critical Asset (First 3 Months) Sarah started with P-401 – their most critical asset, where she fully understood the failure modes through previous RCM analysis. One temperature sensor, local monitoring, manual data analysis. Investment: $1,600, focus on learning and proving the concept.

Phase 2: Expansion to comparable assets (months 4-8) After proven success, Sarah implemented identical sensor solutions on all critical centrifugal pumps. Five assets, standardized procedure, basic cloud integration. Investment: $6,000, focus on scaling and automation.

Phase 3: Advanced Analytics (months 9–15) With sufficient data, Sarah experimented with AI algorithms for pattern recognition. Machine learning software, extensive data analysis, predictive models. Investment: $8,000, focus on insight and optimization.

Decision Framework for Digital Investments

Sarah developed criteria for evaluating digital technologies:

Technical suitability: Is there a proven sensor for the parameter to be measured? Does the technology function reliably in the operational environment?

Economic justification: Are the costs lower than the avoided damage? What ROI can we realistically expect?

Operational integration: Does it fit our current processes? Do we have the expertise for implementation and maintenance?

Strategic value: Does it strengthen our RCM approach? Does it deliver competitive advantage?

This decision framework prevents technology-oriented investments that deliver no practical value.

10.6. Balance Between People and Technology

Technology as a Tool, Not a Replacement

Sarah's team still consists of the same people, but they work more effectively. Technicians get better information for maintenance decisions through real-time sensor data. Planners create more reliable schedules based on predictive analytics. Managers gain insight into asset performance through automated reporting.

"Our expertise is enhanced, not replaced," Sarah said. "Sensors detect anomalies, but we decide what to do with them. AI predicts problems, but we determine the priorities. Cloud platforms analyze data, but we interpret the results."

Development of Digital Competencies

Digital transformation requires new skills from maintenance organizations. Sarah's team had to learn to work with sensor data, dashboard interpretation, and predictive analytics. Some team members were enthusiastic, others initially skeptical.

"Training was more important than we thought," Sarah said. "Not just technical training, but also change management. People had to learn to trust data alongside their experience."

Sarah organizes monthly sessions where the team discusses sensor results and identifies areas for improvement. This combination of technology and teamwork strengthens both digital competencies and mutual collaboration.

10.7. Realistic Expectations and Pitfalls

What Digitalization Can and Cannot Do

Sarah's experience provides realistic insights into the possibilities and limitations of digital maintenance technologies.

What it does do: Earlier detection of problems, more accurate predictions, more efficient resource planning, and better documentation of **asset performance.**

What it does not do: Solve fundamental maintenance problems without a systematic approach, predict failure modes that have never occurred, replace decision-making with automation, and reduce costs without investment.

Common Implementation Pitfalls

Sarah identifies typical pitfalls that organizations can avoid:

Too-high expectations: "Our supplier promised 95% accuracy in failure prediction. In reality, we achieve 70%, which is valuable but not revolutionary."

Underestimated complexity: Installing sensors is relatively straightforward, but data integration and organizational change require more time than planned.

Technology-first approach: Organizations that begin with advanced AI without foundational maintenance processes experience disappointment and wasted investments.

Insufficient change management: Technical implementation succeeds, but organizational acceptance fails due to a lack of training and communication.

10.8. The Road Ahead: Continuous Learning and Improvement

Evolution, Not Revolution

Sarah's digital transformation shows that successful implementation of maintenance technologies is an evolutionary process that builds on solid foundations. Organizations that add new capabilities step by step to proven methodologies achieve sustainable results.

"We're nowhere near finished," Sarah said. "New sensor technologies are becoming available, AI algorithms are getting smarter, integration possibilities are improving. But we experiment carefully and always build on what has been proven to work."

Future Developments

Sarah monitors several technological developments: augmented reality for maintenance instructions, blockchain for parts authentication, digital twins for complex system simulation, and edge computing for faster local data processing.

"We look at everything, but only try what fits our systematic approach," Sarah said. "Technology must strengthen our RCM methodology, not replace it."

Advice for Organizations

Sarah's recommendations for organizations considering digital maintenance technologies:

"First get your foundation in order with systematic preventive maintenance. Start small with proven technologies on critical assets. Invest as much in people change as in technology. Measure results and learn from experience. Be patient — digital transformation takes years, not months. The best part of our digital journey is that it has made our team stronger," Sarah said. "We have better tools, but more importantly: we think more systematically, work data-driven, and remain curious about improvements."

The combination of proven RCM methodologies and smart digital enhancement offers organizations a powerful route to world-class maintenance. Start with a systematic approach, add technology where it delivers value, and keep focusing on the real goal: reliable assets that give your organization a competitive advantage.

Epilogue: Preventive Maintenance in Perspective

Sarah, Eighteen Months Later

Sarah walked through the plant she had come to know a year and a half earlier as a collection of crises and fires. Now the dashboard showed numbers that made the difference: 40% less unplanned downtime, 25% lower maintenance costs.

"Nine times out of ten when my phone rings, it's a planned work order instead of a crisis," she said.

But the eighteen months had also brought new challenges. Work orders that were executed incorrectly due to unclear instructions. Parts that were not in stock. Assembly errors during the preventive maintenance itself.

"Systematic preventive maintenance solves a lot," Sarah acknowledged. "But I also saw where it falls short."

What Preventive Maintenance Does and Does Not Solve

Sarah's transformation demonstrates the power of systematic preventive maintenance. You now know which assets are critical, what maintenance they require, and why certain strategies work.

But as Sarah discovered, challenges remain:

Work Organization: Planning, prioritizing, and executing thousands of work orders requires a systematic approach that goes beyond knowing what needs to be done.

Recurring problems: Some failures keep repeating despite preventive maintenance, because the underlying design flaw or process deviation is not being addressed.

Inventory management: Sensors reliably predict when parts need to be replaced, but if those parts have a four-week lead time while you only have two weeks...

Assembly and execution: Even perfect maintenance fails if the execution is incorrect. Improper alignment, incorrect torque values, or contamination during assembly will undermine your preventive strategy.

Different Routes to the Same Goal

Sarah began, as many organizations do, with preventive maintenance. It delivered quick, measurable results and created support for further improvements. But it is not the only starting point.

Some organizations start with autonomous maintenance — operators taking basic care of their machines. This works especially well in companies with motivated operators and simple machines. It creates engagement and ownership, which facilitates later improvements.
Others start with work flow control because their maintenance department is drowning in chaos and ad-hoc work. Or with structural failure reduction because one recurring problem is their biggest cost driver.

The Nine Disciplines for Structurally Reliable Assets

The challenges that remain — from work organization to troubleshooting, from inventory management to design flaws — each call for their own systematic approach. This book series covers nine disciplines that together answer the question: "What must an organization do to make its technical assets perform as well as possible?"

1. Maintenance Management - The Overarching Framework
2. Planning and Scheduling in Maintenance- Efficient Organization of Work Activities
3. Troubleshooting Breakdowns- Effective Intervention in Acute Problems
4. Root Cause Analysis in Maintenance- Addressing Recurring Problems
5. Preventive Maintenance - Proactive Asset Care
6. Spare Parts Management - Optimal Inventory Control
7. Precision Maintenance- Flawless Installation and Overhaul
8. Engineering for Reliability - Reliability During Design
9. Autonomous Maintenance - Operator Involvement

The nine books follow a logical structure in which each book builds on knowledge from earlier parts. The implementation order in your organization may differ, however — start where the pain is greatest and where you can achieve results quickly. In the introduction of this book you will find a complete overview of all parts and how they relate to one another.

More information about the complete series
For a current overview of all books, additional information and ordering options, visit:

vadeo.nl/books.html

Every discipline you add makes the others more effective. They reinforce each other. Good work flow control ensures you have time for precision work. Preventive maintenance works better when operators detect deviations early. Reliability engineering prevents problems that you would otherwise have to address structurally. The order in which you implement them depends on what your organization needs most.

Your Next Step

You have now learned a methodology that works. TechPro's results—and the thousands of organizations applying similar approaches—prove that systematic preventive maintenance transforms organizations.
The question is: do you stop here, or do you build further?
For now: start with what you have learned in this book. Apply the methodology. Measure your results. Celebrate your successes.
Sarah's advice: "Start where the pain is. Measure results. And if you discover that other aspects are holding you back, there are systematic solutions for those too. We started with preventive maintenance and that

worked for us. But I know companies that started with autonomous maintenance and were just as successful. It's not about the order, but about the systematic approach."

Figures

Glossary — Terminology and Acronyms

AI / Machine Learning Algorithms that recognize patterns in sensor data for failure prediction.

Asset A physical asset that creates value for the organization.

Asset Register (Master Equipment List) A complete and up-to-date list of all assets in an organization with identification codes and basic characteristics.

ATEX European directives for safe working in explosive atmospheres.

Audit Trail A documented history of maintenance activities for compliance purposes.

Availability The percentage of time that an asset is operationally available for production.

Cavitation Damage to pump components caused by the implosion of vapor bubbles in liquid.

Circular Economy An economic model focused on reuse and life extension instead of replacement.

Cloud Platform External server infrastructure for storage and analysis of large volumes of maintenance data.

CO_2 Footprint Total greenhouse gas emissions caused by the production and operation of assets.

Condition Monitoring Systematic monitoring of asset condition through measurement of parameters such as vibrations, temperature, and oil quality.

Condition-Based Maintenance (CBM) Maintenance based on the actual measured condition via inspection or monitoring, so that intervention only takes place when necessary.

Corrosion Chemical or electrochemical attack on materials through reaction with their environment.

Cost-Benefit Analysis A systematic evaluation of preventive maintenance costs versus expected failure costs.

Creep Slow permanent deformation of materials under constant load at high temperature.

Criticality Assessment A multidisciplinary team session in which all assets are systematically evaluated for the impact of failure.

Criticality Ranking A systematic assessment of assets against predefined criteria (safety, economics, environment, quality) to establish priorities.

Digital Twin A virtual replica of an asset or system for simulation and analysis without physical intervention.

EAM (Enterprise Asset Management) Software for planning, execution, and registration of maintenance and spare parts management.

Energy Efficiency The degree to which assets convert energy into useful work; good maintenance increases efficiency by 10–15%.

Facilitator An independent facilitator of RCM sessions who guides the process and keeps the team focused.

Failure Mode The specific way in which a component or system can fail to perform its function.

Failure Pattern The characteristic development of the probability of failure over time; six different patterns (A–F) determine which maintenance strategy is most appropriate.

Fatigue Gradual material damage caused by repeated alternating loads, leading to cracks and ultimately fracture.

Function Test A periodic test to verify whether hidden failures, especially in safety systems, have already occurred.

Functional Failure A situation in which an asset no longer meets the defined functional requirements and performance standards.

HACCP A food safety system with specific maintenance requirements in the food industry.

Industrial RCM (iRCM) A streamlined RCM process in which one engineer does the preparatory work and a team validates; suitable for important but not extremely critical assets.

IoT (Internet of Things) A network of sensors that continuously collect data on asset condition for real-time monitoring.

ISO 55000 International standards series for asset management with requirements for maintenance systems.

Life Cycle Costs (LCC) The total costs of an asset over its entire lifespan, including acquisition, maintenance, and decommissioning.

Load Factor The percentage of maximum capacity at which an asset is loaded; reducing it often increases service life exponentially.

Maintenance Manager Final authority for maintenance strategy and organization.

Maintenance Supervisor Direct supervisor of maintenance technicians in daily operations.

MTBF (Mean Time Between Failures) The average time between consecutive failures of an asset.

MTTR (Mean Time To Repair) The average time required for complete failure resolution from notification to restoration.

NDT (Non-Destructive Testing) Non-destructive testing methods such as ultrasonic, radiography, and penetrant liquid testing for detection of internal defects.

OEE (Overall Equipment Effectiveness) A performance measurement consisting of Availability × Performance × Quality, expressed as a percentage.

Oil Analysis Laboratory examination of lubricating oil for the detection of wear particles and degradation of oil quality.

Payback Period The time required to recover an investment through annual savings.

P-F Interval The time span between the point at which a potential failure becomes detectable (P) and the point at which actual failure occurs (F).

Pilot Project A limited initial implementation to test the approach and learn lessons for broader rollout.

Planning A process in which maintenance activities are prepared with all required information, materials, and safety requirements.

Predictive Maintenance An advanced maintenance strategy in which data analysis and AI predict failures before they occur.

Preventive Maintenance Planned maintenance actions performed before failures occur, with the goal of increasing reliability and preventing unplanned downtime.

Quick Wins Simple improvements that can be implemented quickly for immediate, visible results.

RCM (Reliability Centered Maintenance) A systematic methodology for developing maintenance strategies based on seven core questions about functions, failures, and consequences of failure.

Reliability The probability that an asset will meet all specified requirements without failure during a defined period of time under specific conditions.

Reliability Engineer An engineer responsible for systematically developing maintenance strategies and structurally improving technical reliability.

Risk Matrix A tool with severity levels and risk types that determines the maximum acceptable frequency of incidents.

Root Cause Analysis (RCA) A systematic methodology for identifying and eliminating the fundamental causes of failures, rather than treating symptoms. The goal is to resolve problems permanently so they do not recur.

Run-to-Failure Maintenance (RTF) A deliberate strategy to allow an asset to fail and only then repair it, applied when prevention is not economically justified.

Scheduling The scheduler combines prepared work orders into an efficient weekly schedule with optimal resource utilization.

Thermography The use of infrared cameras for visualization of temperature patterns and detection of hotspots.

Uptime The percentage of time that an asset is available for production.

Usage-Based Maintenance (UBM) Periodic replacement or overhaul of components based on a fixed interval (time, operating hours, miles), regardless of actual condition.

Vibration Analysis Measurement and analysis of vibrations on bearing housings for detection of imbalance, misalignment, and bearing damage.

Wear Material loss due to mechanical interaction between moving surfaces.

Weibull Distribution A statistical distribution that describes the service life variation of components and helps determine optimal replacement intervals.

Wöhler Curve (S-N Curve) A graph showing the relationship between stress and number of cycles to failure due to fatigue.

Work Order A formalized assignment containing all information for the execution of maintenance activities, including materials, instructions, and safety requirements.

Work Preparation A process in which maintenance activities are prepared with all required information, materials, and safety requirements.

Acronyms and Abbreviations

AI Artificial Intelligence

ATEX ATmosphères EXplosibles

CBM Condition-Based Maintenance

CO_2 Carbon dioxide

EAM Enterprise Asset Management

FMEA Failure Mode and Effect Analysis

HACCP Hazard Analysis Critical Control Points

IoT Internet of Things

iRCM Industrial Reliability Centered Maintenance

ISO International Organization for Standardization

KPI Key Performance Indicator

MTBF Mean Time Between Failures

MTTR Mean Time To Repair

NDT Non-Destructive Testing

OEE Overall Equipment Effectiveness

P&ID Piping and Instrumentation Diagram

RCA Root Cause Analysis

RCM Reliability Centered Maintenance

ROI Return on Investment

TPM Total Productive Maintenance

UBM Usage-Based Maintenance

www.ingramcontent.com/pod-product-compliance
Lightning Source LLC
Chambersburg PA
CBHW050528240926
51846CB00003B/739
9789083643168